"I think that modern physics has definitely decided in favour of Plato. In fact the smallest units of matter are not physical objects in the ordinary sense; they are forms, ideas which can be expressed unambiguously only in mathematical language."

Werner Heisenberg (1901-1976)

Published by Our Publishing
London, UK

www.intwoinfinity.com

THE TEMPLATE

A Journey through 3D

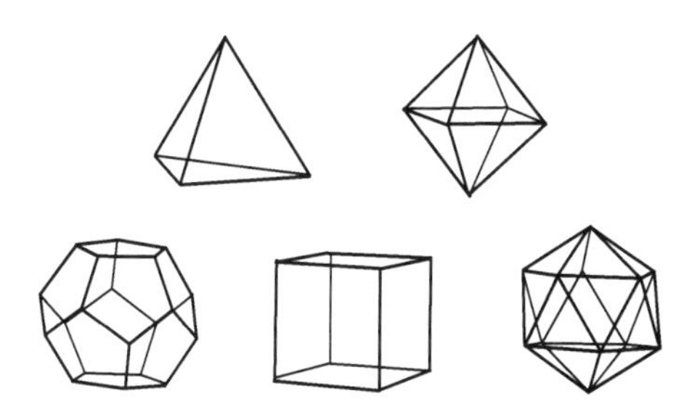

written by

William W. Williams

edited & illustrated by

Heike Bielek & Colin Power

"Then God blessed the seventh day and made it holy."

Genesis 2:3

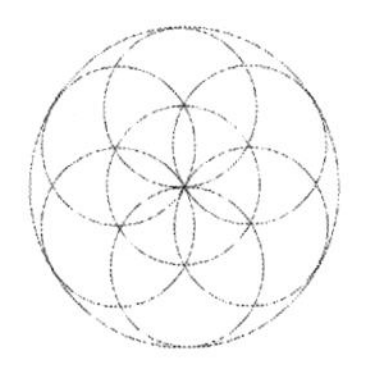

"Once the form was complete,
the time for contemplation arrived."

William W. Williams

CONTENTS

WELCOME GEOMETRICIANS
introduction into sacred geometry

Welcome to this practical introduction into the art of 3D Sacred Geometry. This little book is a guide for the creation of the five *Platonic Solids* using only a drawing compass and a straight edge.

Each solid arises from its respective 2D template, or net (*opposite*). These are folded to complete a 3D polyhedron.

The term '*sacred*' indicates that these forms are all created without measuring any distance. All good scientists know, 'everything is relative'. Thus, the radius of the drawing compass determines the size and side length (or edge) of each form.

Additionally, contained within these pages is a wealth of information about how the Platonic Solids relate to each other, and the natural world.

With a little bit of practice and a lot of fun, you can turn a flat piece of paper into a fine looking three dimensional object!

So, let's begin our exploration into the fascinating world of 3D Sacred Geometry.

THE TEMPLATES

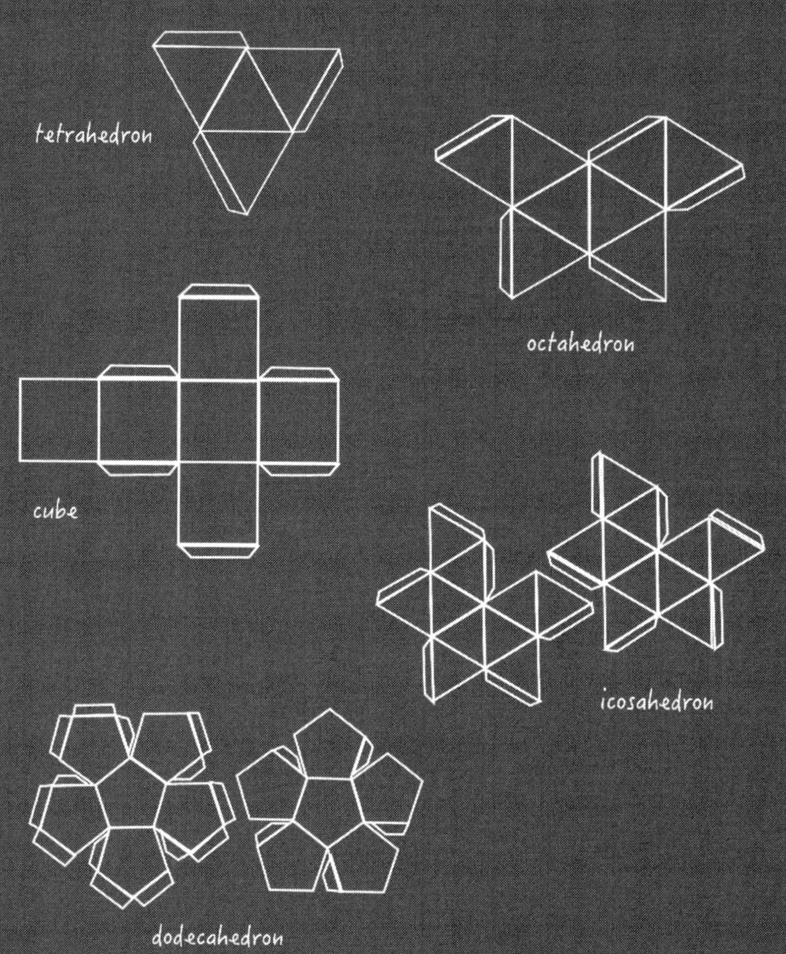

THE PLATONIC SOLIDS

overview

There are only five Platonic Solids in existence.

1) The Tetrahedron with **4** triangular faces
2) The Octahedron with **8** triangular faces
3) The Cube with **6** square faces
4) The Dodecahedron with **12** pentagonal faces
5) The Icosahedron with **20** triangular faces

What makes these shapes unique is that each face, edge length and all distances from any corner to the centre (*vertex)* are the same. Their faces are comprised of either a triangle, a square, or a pentagon.

The name 'Platonic Solids' derives from the Greek Philosopher Plato (427-347BC), many refer to him as the father of Western Philosophy. However, evidence of their discovery dates back much further than that. A 4000 year old set was discovered at the giant stone circles in Aberdeenshire, Scotland (*below, Oxford museum*).

Philosopher Plato

triangle

square

pentagon

THE FIVE PLATONIC SOLIDS

TETRAHEDRON
4 vertices
4 triangles

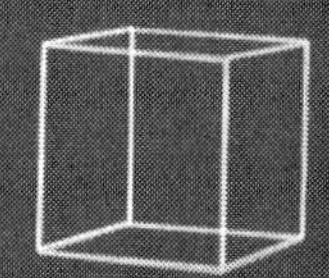

CUBE
8 vertices
6 squares

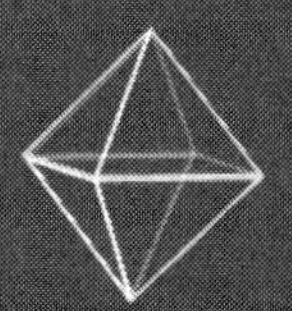

OCTAHEDRON
6 vertices
8 triangles

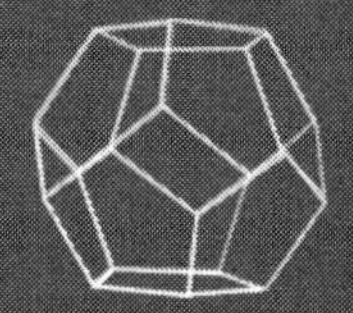

DODECAHEDRON
20 vertices
12 pentagons

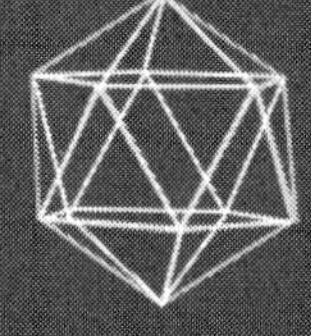

ICOSAHEDRON
12 vertices
20 triangles

THE TOOLKIT
compass, ruler and paper

The Platonic Solids are quite easy to create with the right tools. Here is a checklist of all the things you need:

1) a pencil (and a sharpener)
2) paper (carton paper)
3) a drawing compass (maybe two)
4) paper glue
5) a straight edge (ruler)
6) scissors or cutting knife
7) colour pens, paint (optional)

A word of advice about the compass. There is no substitute for quality. You can use two or more compasses; one set at a fixed distance and another that is flexible. Sometimes we return to one distance after measuring a point of another length. Drawing compasses are of different types, and some have a locking wheel. These are generally smaller and are good as a 'fixed' compass.

For a clear fold, use the ruler to bend the paper along the lines. Before folding, you might like to colour each side. Lastly, a toothpick is handy to help press the tab onto the inside for a clean finish! Can you think up any games to play using them as dice? Or maybe hang them as a set?

TOOLSET

pencil

paper

compass

glue

straight edge/ ruler

scissors

decoration

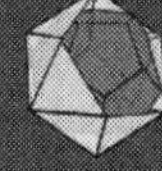

nested

dice

THE SEED OF LIFE
from one to seven circles

Some people consider the *Seed of Life* to be the blueprint of creation. The oldest depiction is etched into the Temple of Osiris in Abydos, Egypt. It is the core element of Sacred Geometry that stands at the beginning of all form. This includes the manifestation of the Platonic Solids.

Let's clarify how this tapestry expands with a compass set at any distance. Whatever distance is set, will define the side length and size of a particular Platonic Solid.

First, create a circle (*1*). Then, place the compass on any point on its circumference to create a second circle. These two circles overlap and create a form called the *Vesica Piscis* (*2*). By centring the compass point over these nodes, you can draw a third (*3*) and fourth circle. This defines a slim petal shape, called the *Trion Re* (*4*). The process continues until seven circles complete the Seed of Life (*5, 6, 7*). The pattern expands to 19 circles forming the *Flower of Life* (*8*).

A total of 61 circles completes the final level of the Flower of Life (*9*). From this image, the *Fruit of Life* emerges, made of 13 full circles (*10*). By connecting all their centres together, we generate Metatron's Cube (*11*), which forms a two dimensional image of all five Platonic Solids (*appendix, page 40*).

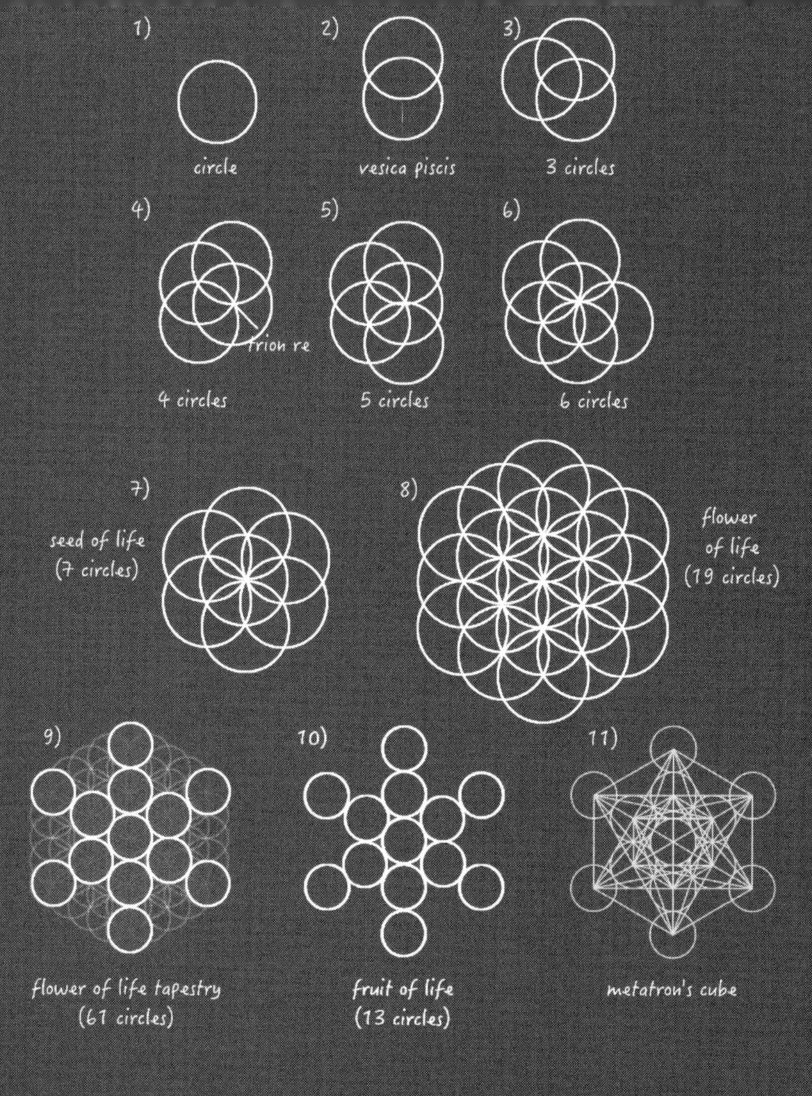
1)
circle
2)
vesica piscis
3)
3 circles
4)
trion re
4 circles
5)
5 circles
6)
6 circles
7)
seed of life
(7 circles)
8)
flower
of life
(19 circles)
9)
flower of life tapestry
(61 circles)
10)
fruit of life
(13 circles)
11)
metatron's cube

DRAWING THE TETRAHEDRON

from three circles

The first three circles of the Seed of Life create the template for the Tetrahedron. With only four faces, it is the most simple of all the Platonic Solids.

Use the compass to draw three circles (*1*). Connect the nodes with the ruler (*2*). This completes the template for the Tetrahedron. Cut out the form and leave tabs as indicated (*3*). Simply, fold each triangle upwards to meet in the centre and glue the tabs together (*opposite middle/bottom*). Congratulations! You just finished the first Platonic form.

The Tetrahedron is the only Platonic Solid that rests with its point facing upwards. All the others sit with a flat face, when placed on a surface (*below*). This feature gives the Tetrahedron unique properties, since it has no direct axis going through its centre from tip to tip.

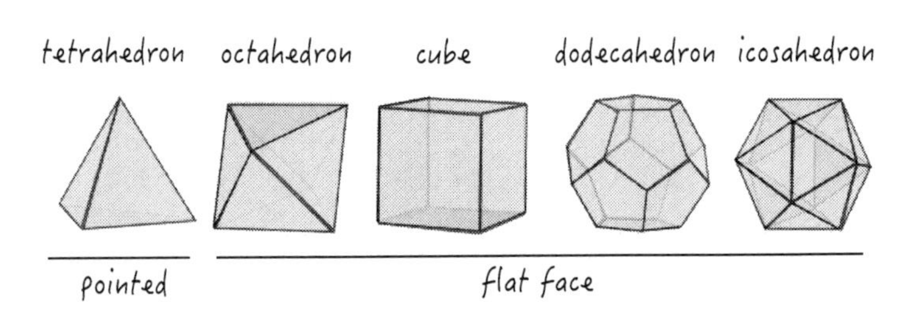

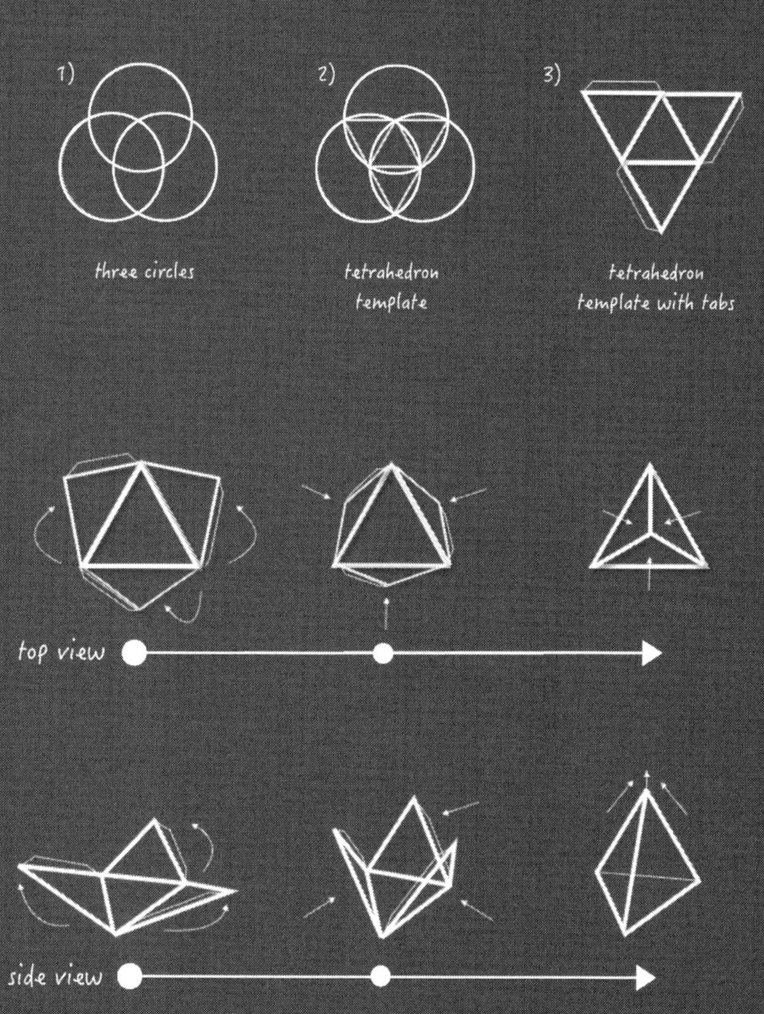
1)
2)
3)
three circles
tetrahedron template
tetrahedron template with tabs
top view
side view

DRAWING THE OCTAHEDRON

from six circles

The Octahedron is also created from the Seed of Life. The process is similar to the Tetrahedron, only this time it begins with the first six circles (*1*). Connect the nodes (*2*) to create the eight triangular sides of the Octahedron (*3*). These are arranged in an inner and outer ring (*4*). When the four triangles in the inner ring fold together, the form bends into 3D. This creates a pyramid with a square base (*5*). The other four triangles fold together to form a point on the opposite side (*6*). This creates two pyramids on top of each other with a square base in the middle. However, there are a total of three squares hidden within the frame of the Octahedron. Can you find them?

Only two types of regular shapes, the square and the triangle, can fill a flat two dimensional plain (*below*). The Octahedron is the only Platonic Solid that inherits both shapes. That makes it the perfect form to house the two types of two dimensional space (*appendix, page 46*).

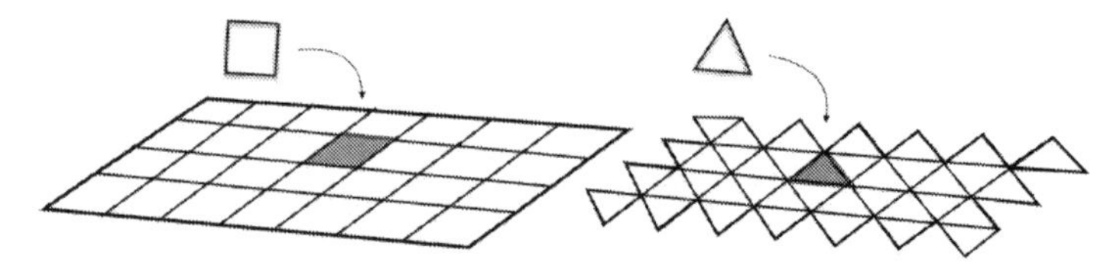

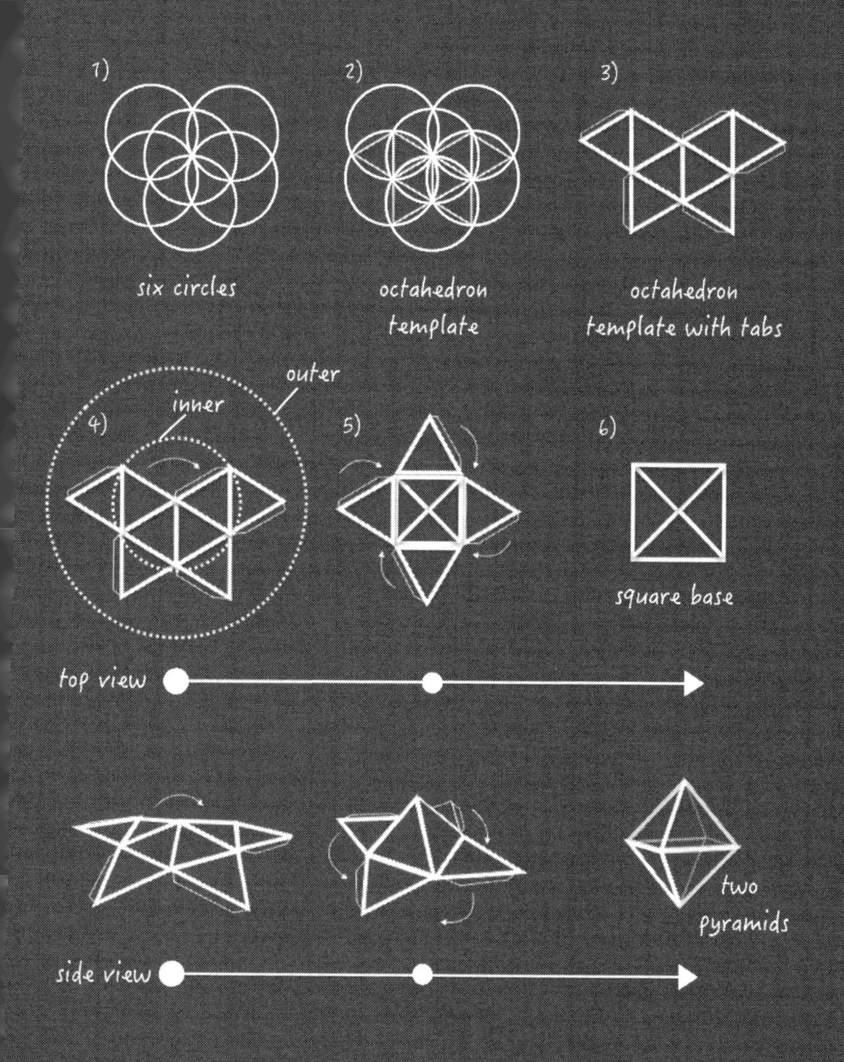
1)
six circles
2)
octahedron
template
3)
octahedron
template with tabs
outer
inner
4)
5)
6)
square base
top view
two
pyramids
side view

DRAWING THE ICOSAHEDRON

from seven circles

The last Platonic form, derived from the Seed of Life, is the Icosahedron. It takes seven circles, one whole seed, to create one half. The full form requires two identical templates with different tabs (*below*).

Draw the Seed of Life (*1*) and connect the nodes (*2*) to create the first half of the template (*3*). The triangles are distributed between an inner and outer ring, similar to the Octahedron (*4*).

This time, the structure bends into three dimensional space (*5*) and forms a pyramid with a pentagonal base (*6*). Two templates interlock with each other to complete the final form (*opposite, bottom right*).

template 1

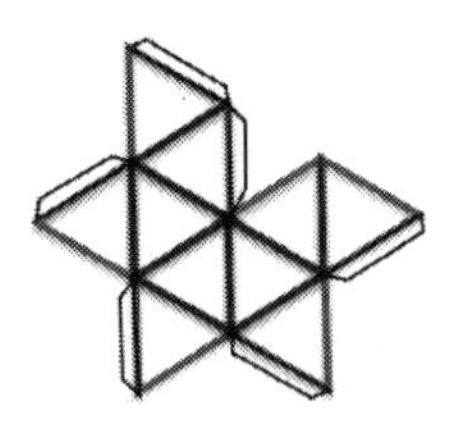

template 2

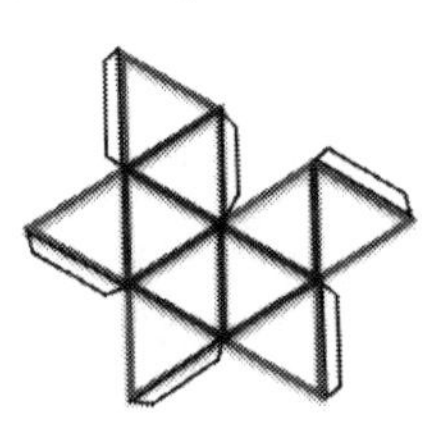

icosahedron

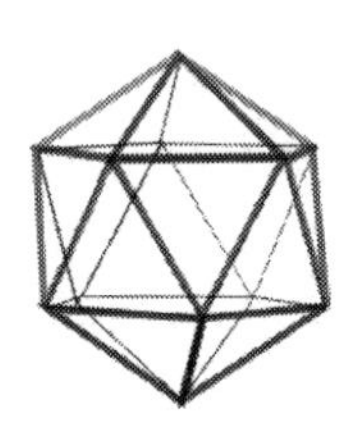

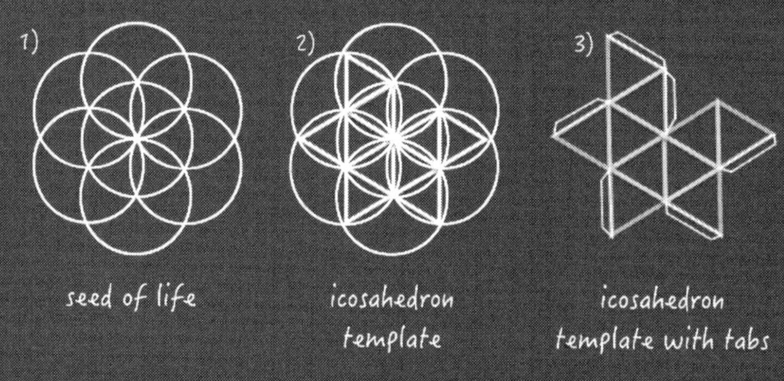
1)
2)
3)
seed of life
icosahedron
template
icosahedron
template with tabs

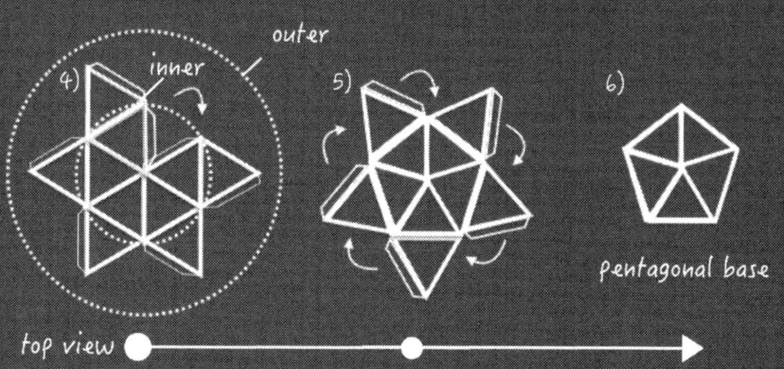
outer
inner
4)
5)
6)
pentagonal base
top view

template 1
template 2
side view

DRAWING THE CUBE

from a circle to a square

The square only uses a single sized circle in its creation. Fixing the compass at one length, begin by drawing three circles of the Seed of Life (*1*). Connect the nodes of the Vesica Piscis, where two circles overlap (*2*). The drawn line crosses the third circle and defines a point on its circumference. Centre the compass on this point and draw a fourth circle (*3*). This defines the four corners of a square (*4*) and the first face of the cube (*5*).

The Seed of Life forms a triangular 2D plane that contains the Trion Re. After four circles the first Trion Re appears (*bottom left*). The circle that marks the top two corners of the square is transposed from the others. It has its centre placed on the top of the Trion Re (*6*). This translates the lines of 2D space from a triangular to a square formation. The Trion Re holds the blueprint for this transformation to occur (*below*).

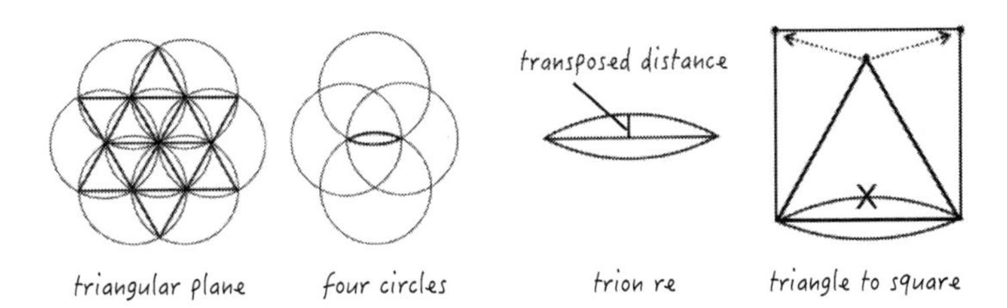

triangular plane — four circles — trion re — triangle to square

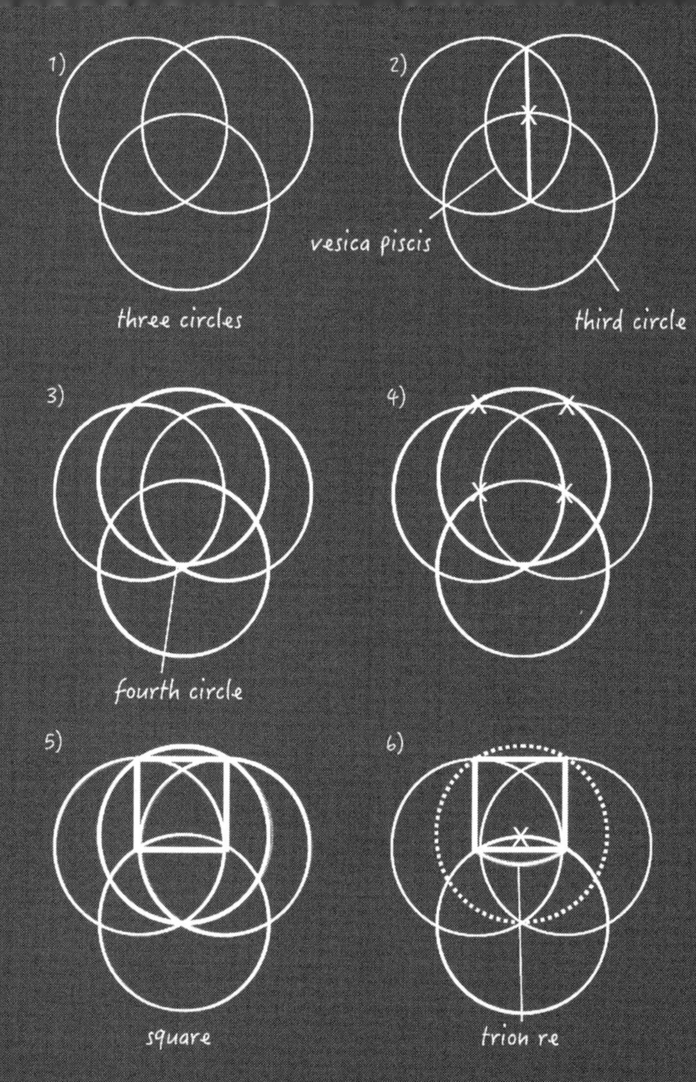
1)
three circles
2)
vesica piscis
third circle
3)
fourth circle
4)
5)
square
6)
trion re

DRAWING THE CUBE

from a square to the cube

Once the square is defined, it is easy to create the template for the Cube. Take your ruler and extend the parallel sides of the square in all directions (*1*). Keeping the compass at the same distance, draw four circles on the corners of each square (*2*). Where each circle crosses the extended lines, another four squares are defined. Use these points as a reference to create a cross made of five squares (*3*). Extend one side of the cross, using the same technique to complete the template (*4*). An alternative template is detailed at the back of this book (*appendix, page 39*).

The Cube is representative of our dimension of space. We generally measure space in terms of length, depth and breadth of a Cube. It is the only Platonic Solid to fill space seamlessly and completely by itself (*appendix, page 46*). This type of solid belongs to the 'Space-Filling Polyhedra'.

cube
"space filling polyhedra"

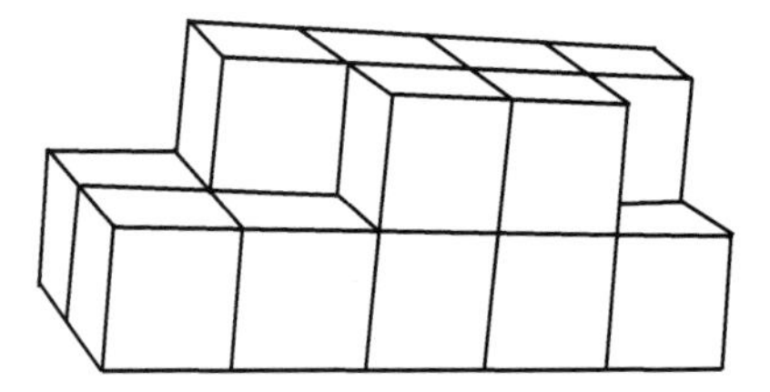

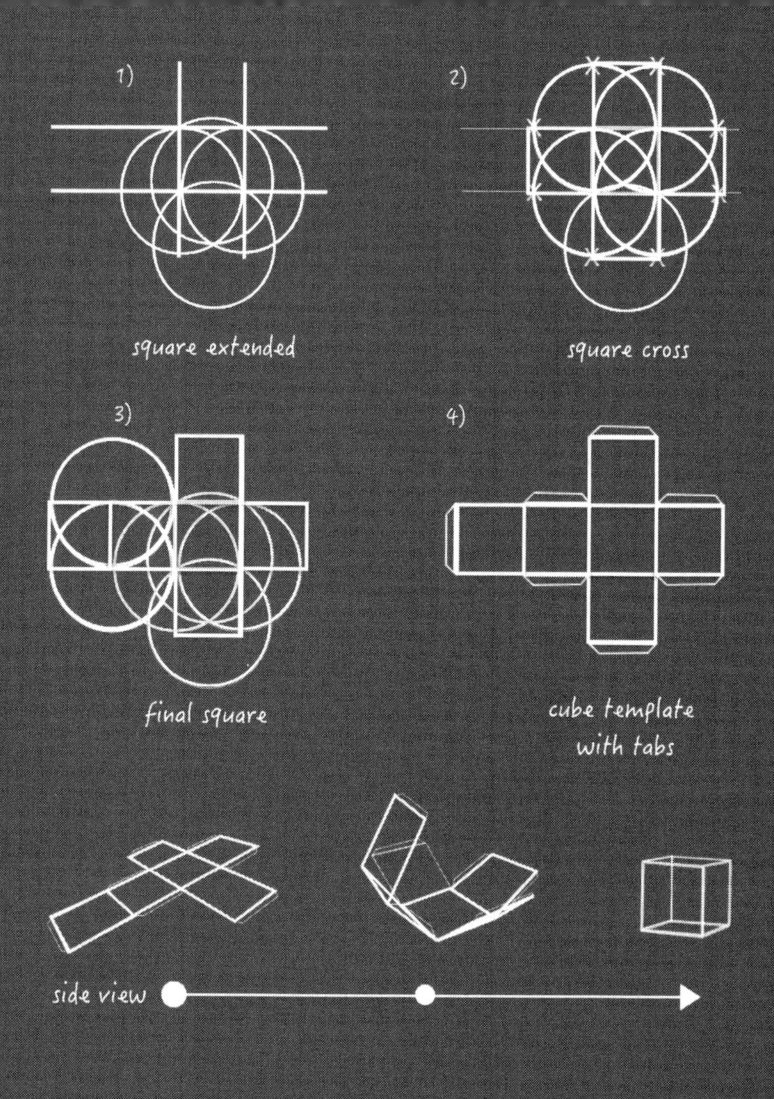

1)
2)
square extended
square cross
3)
4)
final square
cube template
with tabs
side view

DRAWING THE DODECAHEDRON
from a square to a pentagon

The Dodecahedron is the only Platonic Solid with pentagonal faces. The pentagon evolves out of the square (*1*). They both share the same two bottom corners.

To define the tip of the pentagon, connect the nodes of the Vesica Piscis and extend the line (*2*). Adjust the compass to measure the diagonal of the square (*3*). Next, focus the compass on the intersection of the circle and the centre-line, just as we did with the square (*4*). Create a forth, larger circle, which overlaps with the centre-line (*5*). This defines the tip of the pentagon (*6*). Re-adjust the compass back to its original distance and draw a circle centred on the tip of the pentagon (*7*). This crosses the Vesica Piscis to define the last two corners of the pentagon (*8*). Now, you can connect all the nodes to create the first face of the Dodecahedron (*9*).

This is the first time we changed the dimension of the compass. The diagonal of a square measures $\sqrt{2}$ if its side length is 1 (*appendix, page 40*). Again, this larger circle is transposed from the others by a distance defined by the Trion Re. We generated the square in the same way (*page 15*), only this time the radius of the circle has changed from 1 to $\sqrt{2}$ (*opposite, below*).

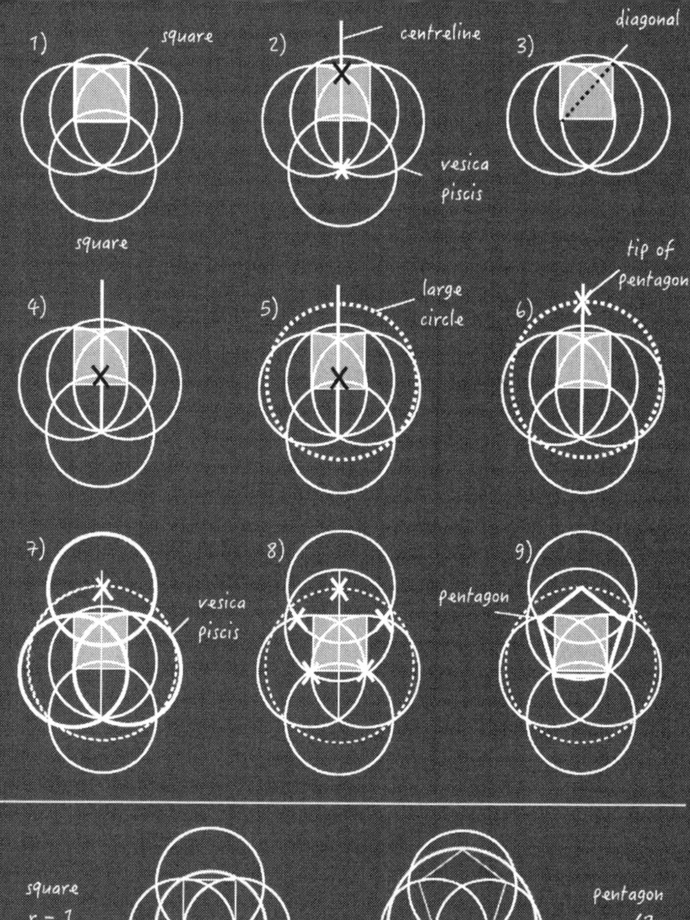

square
r = 1

pentagon
r = √2

DRAWING THE DODECAHEDRON

from a pentagon to the template

We created the full template of the Cube by extending the sides of a square. A similar process applies to the Dodecahedron.

Connect all corners of the pentagon to form a five pointed star and extend these lines outwards (*1*). Set the compass to the side length of the pentagon and draw circles on each corner. Where these cross the extended lines, marks the first two corners of the neighbouring pentagons (*2*). Centre the compass on these points and draw two circles. Where these overlap, defines the tip of each pentagon (*3*). Connect the outer points to create a larger pentagon. This creates the first half of the template (*4*).

Just like the Icosahedron, it takes two identical versions to create the final form (*opposite, bottom*). Be sure to attach the tabs in different places on each template (*below*).

Congratulations, you have now mastered the creation of all five Platonic Solids!

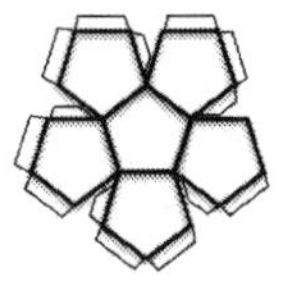

template 1

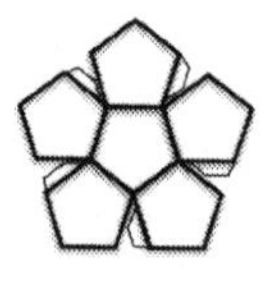

template 2

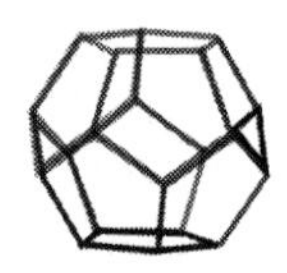

dodecahedron

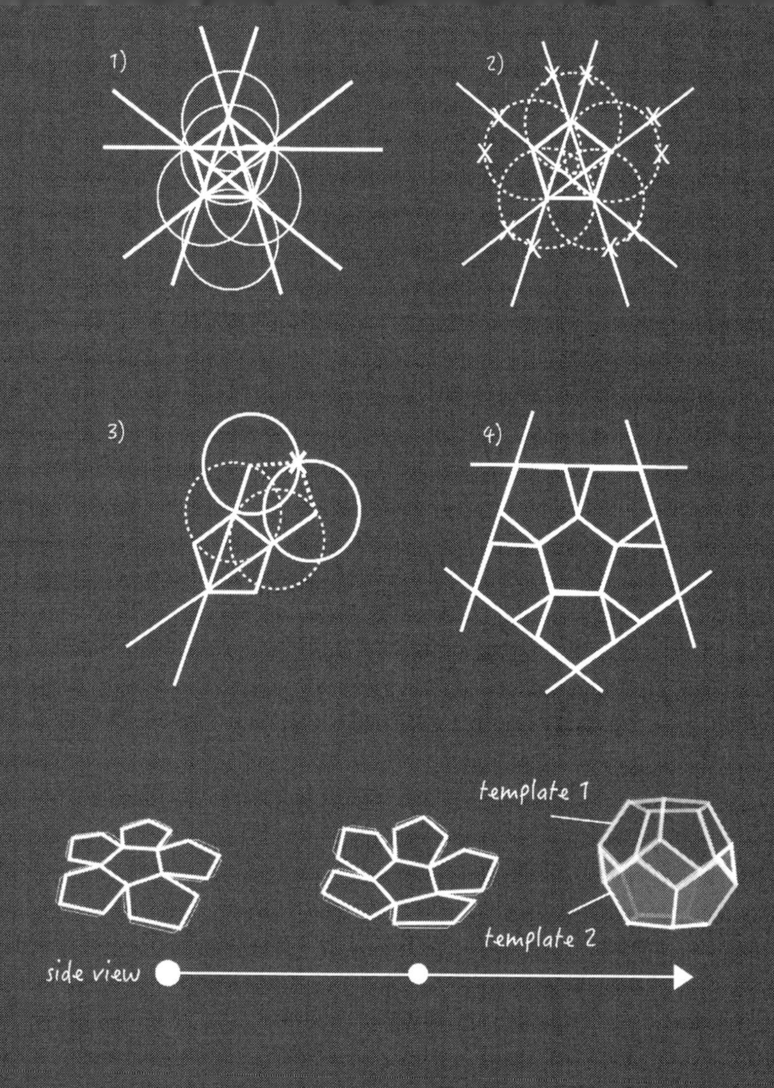
1)
2)
3)
4)
template 1
template 2
side view

NESTING THE SOLIDS

russian dolls

Now that you know how to create the five Platonic Solids, there is still an additional challenge! It is possible to create each solid at a specific size, so they fit perfectly inside one another - just like Russian Dolls. This strengthens each solid, since it gets supported by its neighbouring forms. This is called *Nesting the Solids.*

It occurs in a particular sequence with sides at specific ratios (*1 and appendix, page 41*). Once nested, the five Platonic Solids sit in three different-sized spheres. The smallest sphere defines the Octahedron (*d1*), the largest encases the Icosahedron (*d3).* The central sphere (*d2*) comprises the Tetrahedron, Cube, and Dodecahedron (*2*).

Notice, that each sphere contains only one Solid made from triangular faces, derived directly from the Flower of Life. If the diameter of the 1st sphere is 1, the 2nd sphere will be $\sqrt{3}$ (*3*). This value also appears in the Vesica Picsis (*below*).

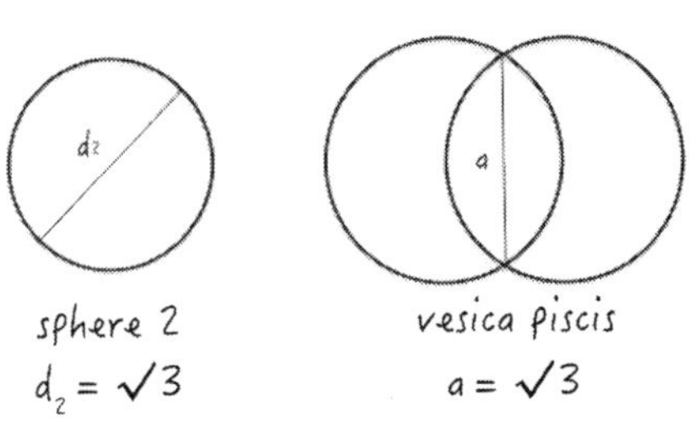

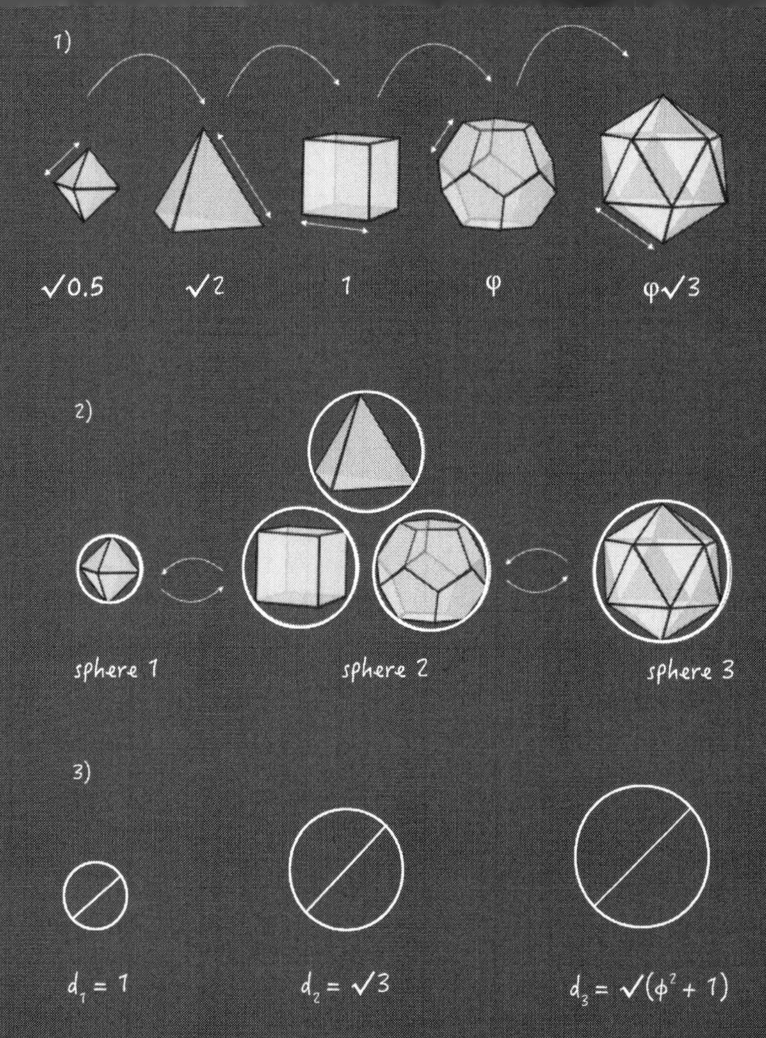
1)
√0.5
√2
1
φ
φ√3
2)
sphere 1
sphere 2
sphere 3
3)
d1 = 1
d2 = √3
d3 = √(φ² + 1)

THE MASTER BLUEPRINT

octahedron, tetrahedron, cube

So far, we have created all the Platonic Solids with no relationship to each other. Now, we need to define the relative sides in a master blueprint in order to nest them properly.

The image on the front of this book is a mandala that incorporates all the necessary lengths in a more artistic manner. For a better understanding, we can simplify this image by breaking it down into its constituent parts (*1*).

Begin by creating the square (*page 15*). Each side represents the edge of the Cube (*2*). Mark a diagonal across the square. This represents the relative edge length of the Tetrahedron (*3*). Draw a second diagonal to divide this length in half. This is the edge of the Octahedron (*4*).

If the side of the Cube measures 1, the side of the Tetrahedron will be $\sqrt{2}$, and the Octahedron will be $\sqrt{0.5}$ (*5*).

With this information the first three Platonic Solids can be nested. At the centre lies the Octahedron, around which fits the Tetrahedron with sides double the length. Both of them fit inside the Cube. Furthermore, the Tetrahedron defines 4 of the Cube's 8 corners (*page 30*).

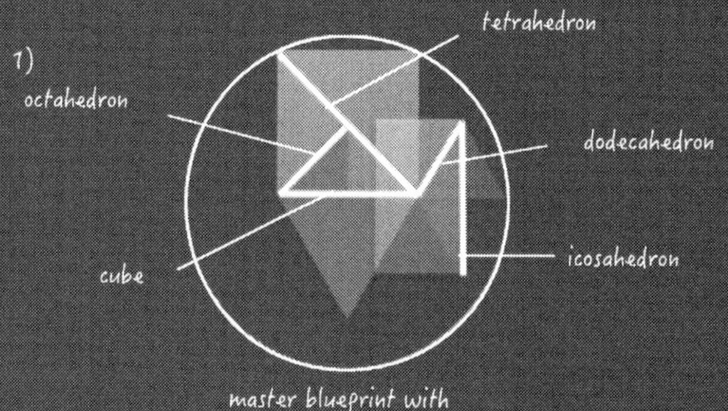

master blueprint with
side length ratios

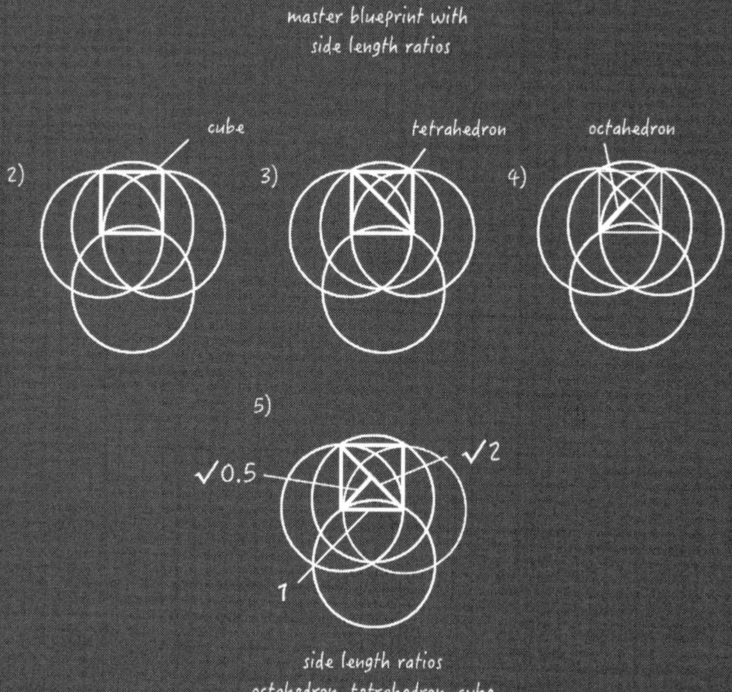

side length ratios
octahedron, tetrahedron, cube

THE MASTER BLUEPRINT

dodecahedron, icosahedron

The next part of the master blueprint focusses on the Dodecahedron and Icosahedron.

Begin with three circles (*1*) and draw lines through each node. This defines the centre point (*2*). Place the compass on this point and adjust the radius to meet the outer nodes. Now draw a larger circle (*3*). At the centre of the three circles sits a small triangle. Extend its sides towards the large circle (*4*). This defines a new distance that is the side length of the Dodecahedron (*5*).

For the Icosahedron, set the compass to the side length of the Dodecahedron (*6*) and draw a Vesica Picsis (*7*). The distance between the two nodes defines the edge length for the Icosahedron (*8*).

With this information of the Master Blueprint, it is easy to create a complete set of nested Platonic Solids.

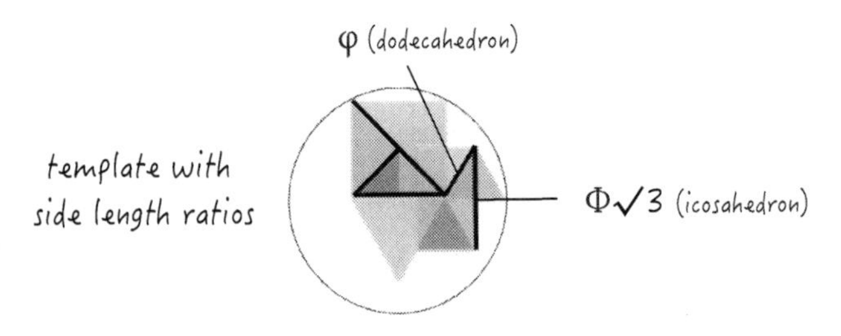

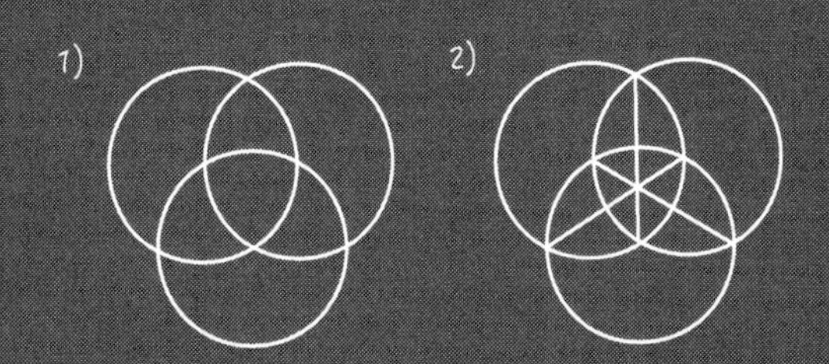
1)
2)

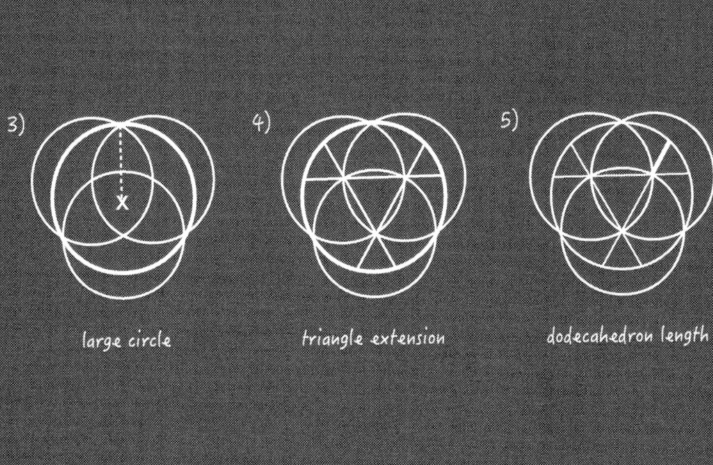
3)
4)
5)
x
large circle
triangle extension
dodecahedron length

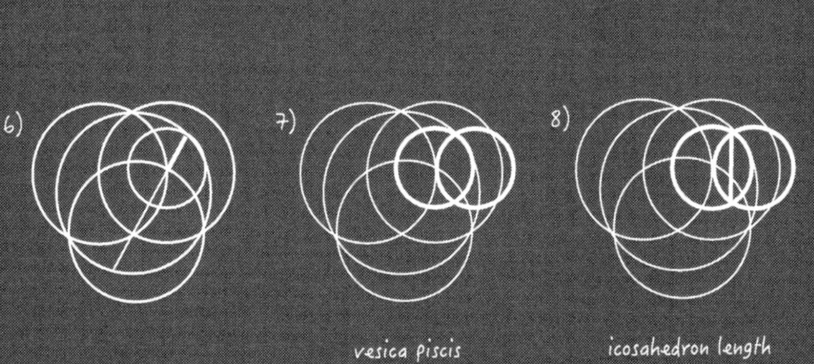
6)
7)
8)
vesica piscis
icosahedron length

NESTING THE OCTAHEDRON

into tetrahedron

The nesting of the Platonic Solids begins with the Octahedron. The master blueprint displays its relative edge length as half the diagonal of a square. If we set the side of the square to 1, then every other length will be relative to that. Now, the Octahedron's side measures $\sqrt{0.5}$.

The next form is the Tetrahedron, with sides of $\sqrt{2}$. This is double the length of the Octahedron, or the full diagonal of a square (*below, left*).

Once nested, notice the relationship between the Tetrahedron and the Octahedron. Each face of the Tetrahedron creates a large triangle. Within this boundary, the Octahedron forms a smaller inverted triangle (*below, right*).

Four small Tetrahedra fill in the gaps between the Octahedron and the large Tetrahedron (*1*). These contain side lengths equal to that of the Octahedron ($\sqrt{0.5}$). Placing four more of these on each face of the Tetrahedron creates a Star-Tetrahedron (*2*). This form defines all 8 corners of the Cube (*3*).

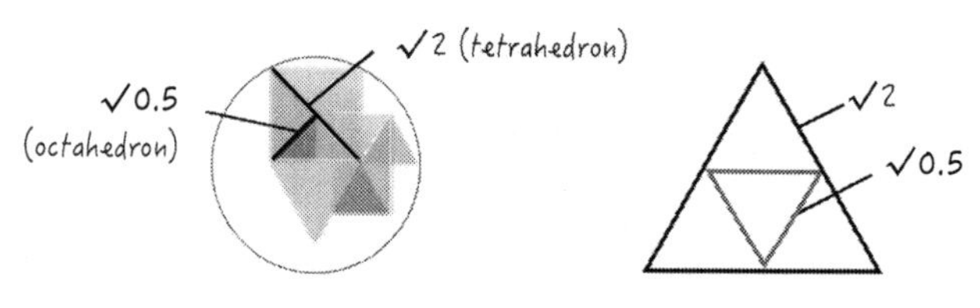

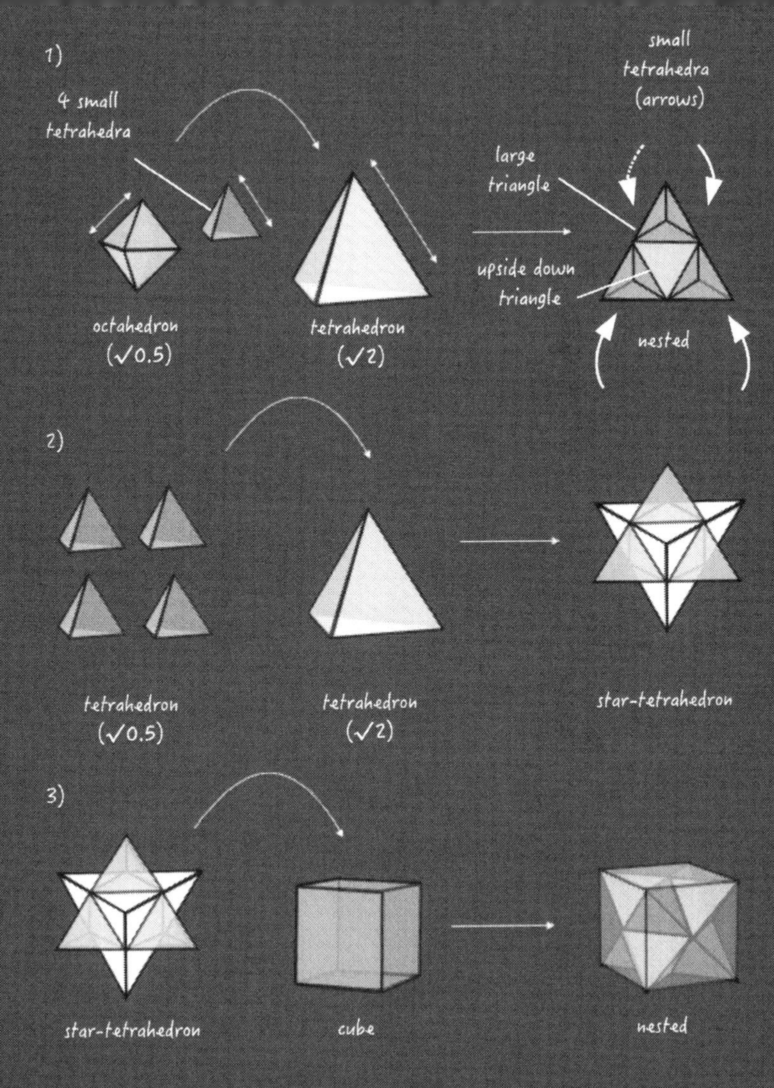
1)
4 small
tetrahedra
octahedron
(√0.5)
tetrahedron
(√2)
small
tetrahedra
(arrows)
large
triangle
upside down
triangle
nested
2)
tetrahedron
(√0.5)
tetrahedron
(√2)
star-tetrahedron
3)
star-tetrahedron
cube
nested

NESTING THE TETRAHEDRON

into cube

The Tetrahedron and the Octahedron comprise of triangular faces. When nested, their sides follow a simple doubling pattern. The sides of the Tetrahedron fall across the diagonal of the square faces of the Cube. This gives the Octahedron a side of $\sqrt{0.5}$, the Tetrahedron $\sqrt{2}$, if the Cube is 1. However, scale is relative. If the side of the Octahedron changes to 1, the sides of the other Platonic Solids will become adjusted relative to that. In this case, the sides of the Tetrahedron will measure 2, and the Cube $\sqrt{2}$ (*below*).

Notice, the Octahedron is the same distance from tip to tip as the edge of the Cube. In this way, the Octahedron defines the centre of each of its square faces (*1*). If the Cube has a side of 1, the diagonal across each face will be $\sqrt{2}$. Therefore, the Tetrahedron with a side length of $\sqrt{2}$ nests perfectly within it (*2*). The furthest distance between two opposite corners of the Cube is $\sqrt{3}$. Thus, the Cube incorporates the ratios $1 : \sqrt{2} : \sqrt{3}$ (*3*).

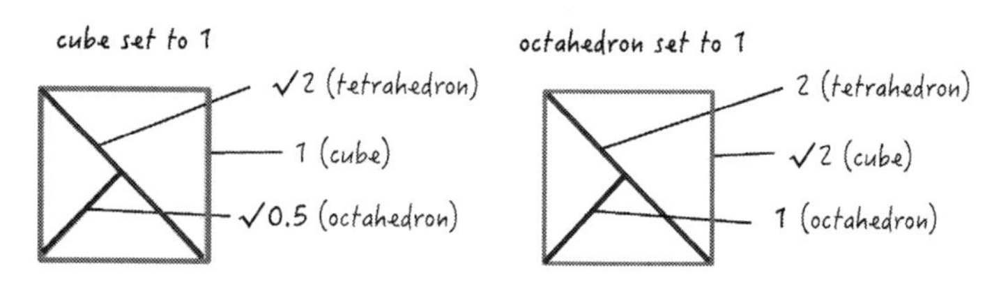

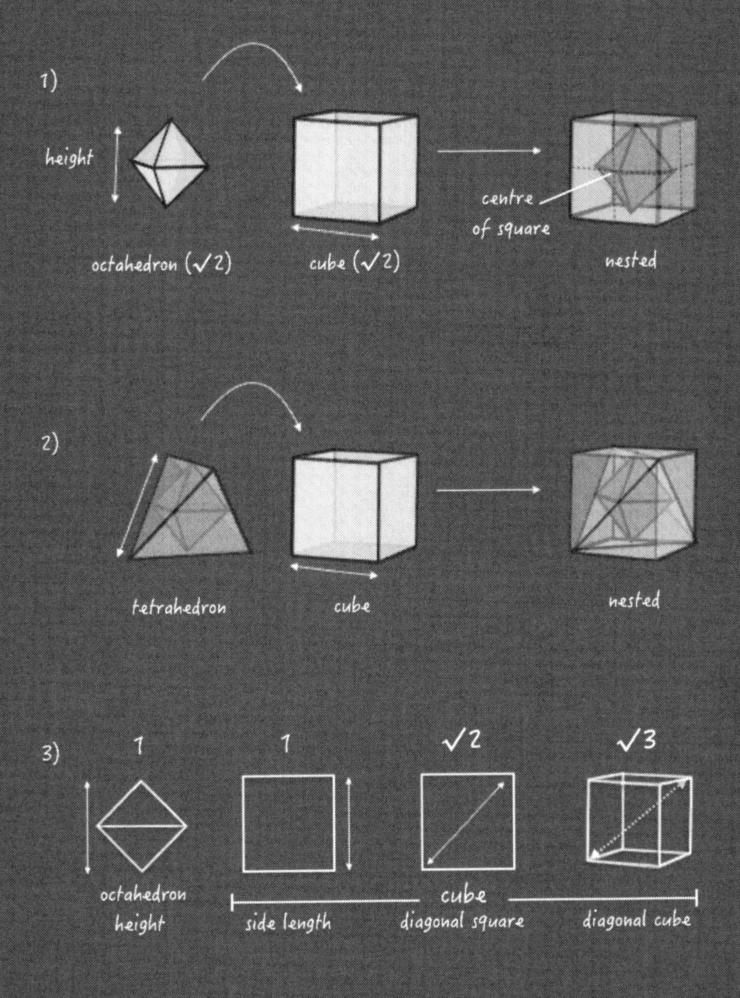

1)
height
octahedron (√2)
cube (√2)
centre
of square
nested
2)
tetrahedron
cube
nested
3)
1
1
√2
√3
octahedron
height
cube
side length
diagonal square
diagonal cube

NESTING THE CUBE
into the dodecahedron

The Cube nests inside the Dodecahedron by incorporating the number *phi*. If the Cube has a side defined as 1, the Dodecahedron will measure phi (φ), the Golden Ratio, or 0.6180 (*1*).

Phi is an irrational number that has two main form, *greater* Phi (Φ = 1.6180...) and *lesser* phi (φ = 0.6180...). We differentiate between these two through a capital "P" or "Φ" and small "p" or "φ".

Due to the unique feature of Phi (*appendix, page 41*), we can invert the ratio of the side length of the Cube and the Dodecahedron. If the Dodecahedron has a side length of 1, the Cube's edges will equal Phi (*2*).

This relationship manifests through the shape of the pentagon. Create a five pointed star by connecting all the corner points to each other. Inside the star a smaller pentagon will form (*3*). If the outer pentagon's side is 1, the internal lines that form the star will measure *greater* Phi. The distance from the corner of the inner pentagon to the outer will be *lesser* phi.

Once nested, the Cube defines 8 of the Dodecahedron's 20 corners. These are also shared by the Star-Tetrahedron (*4*). This completes the second central sphere of the nested set.

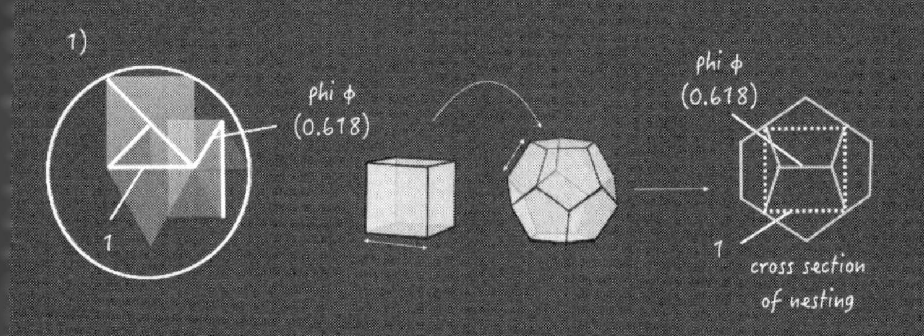
1)
phi ϕ
(0.618)
1
phi ϕ
(0.618)
1
cross section
of nesting

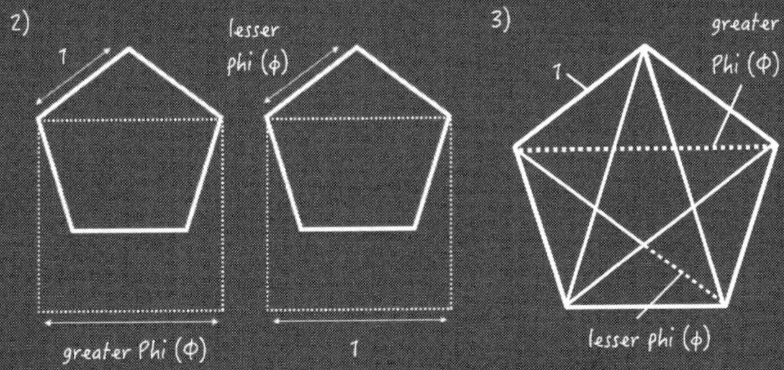
2)
1
lesser
phi (ϕ)
greater Phi (Φ)
1
3)
greater
Phi (Φ)
1
lesser phi (ϕ)

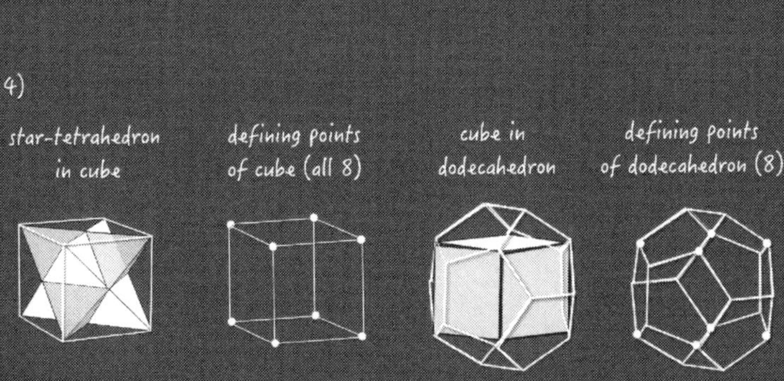
4)
star-tetrahedron
in cube
defining points
of cube (all 8)
cube in
dodecahedron
defining points
of dodecahedron (8)

NESTING THE DODECAHEDRON
into icosahedron

Outside the Dodecahedron nests the final form, the Icosahedron. If the side of the Dodecahedron is 1, the side of the Icosahedron will be $\sqrt{3}$. If the edges of the Dodecahedron measure φ (0.618), the Icosahedron will be $\varphi \times \sqrt{3}$. Nothing changes, only the scale (*1*).

They interface with each other in a similar way to the Octahedron and the Cube. In this case, the Dodecahedron defines the centre of each triangular face of the Icosahedron (*2*). This characteristic is particular to pairs of 'Platonic Duals' (*appendix, page 44*).

The 'top' of the Icosahedron consists of five triangles that fold over to form a pyramid with a pentagonal base. Notice, the pentagon contains the phi ratio, that is found externally on the faces of the Dodecahedron, and internally within the Icosahedron (*3*). Additionally, three interlocking rectangles, with sides measuring phi and one, define the 12 corners of the Icosahedron (*4*). Similar internal geometries apply to the other polyhedra (*appendix, page 48*).

This completes the set of nested Platonic Solids. With the Cube set to 1, the only other side-length ratios used are $\sqrt{0.5}$, $\sqrt{2}$, φ, and $\sqrt{3}$.

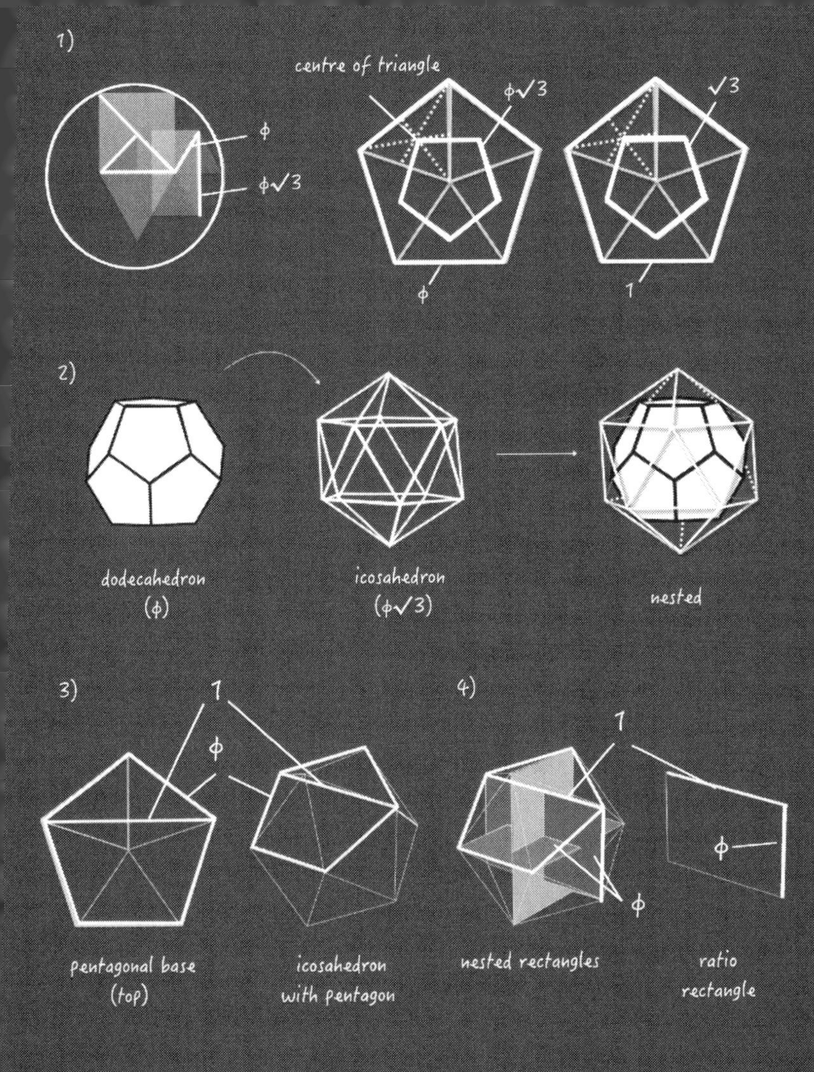
1)
centre of triangle
ϕ
ϕ√3
ϕ√3
√3
ϕ
1
2)
dodecahedron
(ϕ)
icosahedron
(ϕ√3)
nested
3)
1
ϕ
pentagonal base
(top)
icosahedron
with pentagon
4)
1
ϕ
ϕ
nested rectangles
ratio
rectangle

MORE GEOMETRY

the journey is only beginning

Congratulations, you completed the first steps of the metaphysical journey. It started with the circle, forming the Seed of Life, from which the five Platonic Solids manifested.

Additionally, there are another 13 Archimedean Solids, which are beyond the scope of this book (*appendix, page 51*). Two of these - the Truncated Octahedron and the Cuboctahedron - appear in the appendix with their various qualities noted (*appendix, page 52*). Moreover, the appendix reveals a wealth of information about how these shapes relate to mathematics, physics, chemistry, and metaphysics. Their relevance was studied by Johannes Kepler, who used the Platonic Solids to describe the Solar System (*1*). The Golden Spiral is created by the triangles found in the five-pointed star (*2*), which forms the blueprint of many natural forms from animals, plants, shells and even galaxies (*3*). This is just the beginning of an exploration into the depths of reality and the fabric of life itself. And with that thought we leave you with these final words from the great Philosopher himself:

"The direction in which education starts
a man will determine his future in life."

Plato

1)

Kepler's Platonic Solid model of the Solar System from Mysterium Cosmographicum (1596)

detailed view of the inner sphere

2)

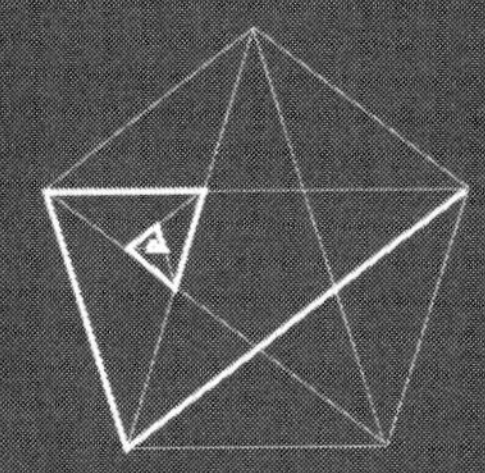

3)

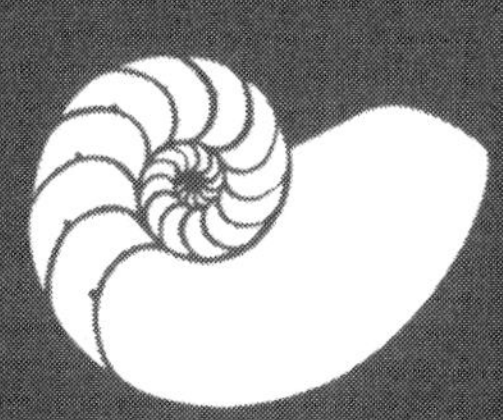

APPENDIX

ALTERNATIVES

Most of the Platonic Solids in this book expand from one face to the full template.
Alternatively, through division we can create a template half the size. To construct the Tetrahedron in this way, begin with three circles (*1*). Draw the sides of the inner triangle (*2*). Connect the nodes to divide each side of the triangle in half (*3*). Connect the midpoint of each side to each other (*4*). Cut out the template and fold into the half sized Tetrahedron (*5*). Four Tetrahedra can be produced from one template. Add these to the faces of a larger Tetrahedron to form a Star-Tetrahedron perfectly (*page 41*). More triangles can be added inside, diminishing in size (*6*). As this process carries on into infinity it forms a fractal.

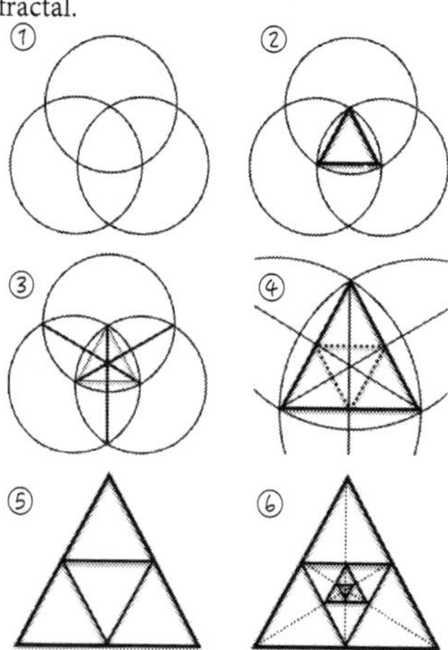

The same process applies to the Cube. In this case, the template is created in two halves - similar to the Dodecahedron and Icosahedron. Start by drawing the square (*page 15*) (*1*). Place the compass on the upper node of the Vesica Piscis to create the Trion Re (*2*). On the opposite side of the Trion Re position the compass to define a square below (*3*). This creates a rectangle made out of two squares (*4*). Use the same radius to draw a circle on each corner (*5*). Connect lines between opposite nodes of the new circles (*6*).

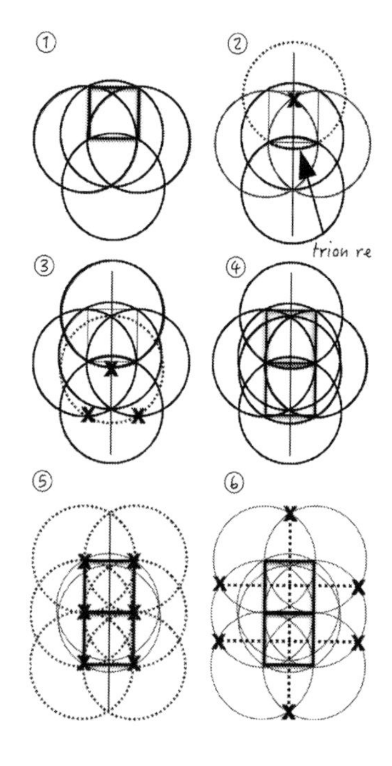

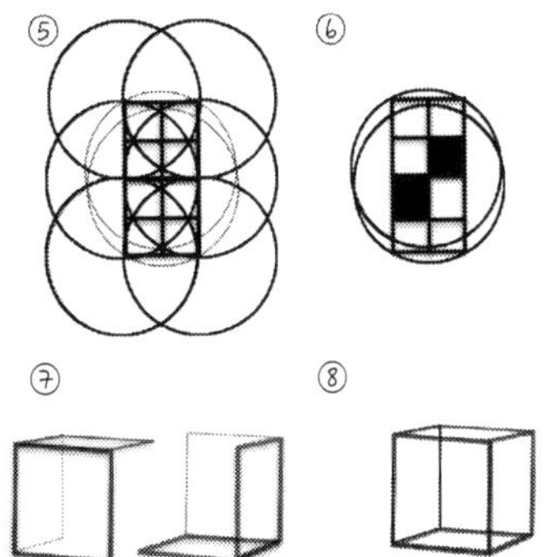

This divides both squares into four smaller ones (*5*). Remove one square from each side (*6*). Cut out and fold the remaining three squares to form three sides of a Cube (*7*). The two halves interlock with each other to complete the form (*8*).

METATRON'S CUBE

Once the Flower of Life expands to a total of 61 circles, it creates the foundation for another form called the Fruit of Life (*page 8*). 13 whole circles are drawn out of this image, and their centres are connected to each other. This forms a design called Metatron's Cube, which creates a 2D projection of all of the five Platonic Solids as well as the Star-Tetrahedron.

1: Octahedron
2: Tetrahedron/Star-Tetrahedron
3: Cube
4: Dodecahedron
5: Icosahedron

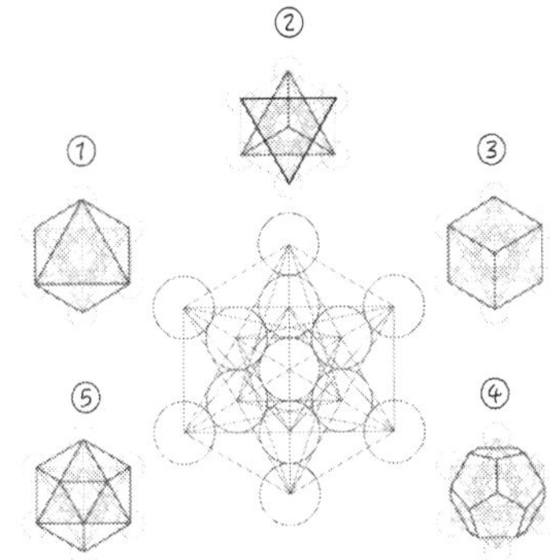

RATIOS

The Platonic Solids nest through a series of ratios. The ratio of $\sqrt{2}$ defines the diagonal of a square. $\sqrt{3}$ measures the distance between two equilateral triangles. The same distance is also found in the Vesica Piscis.

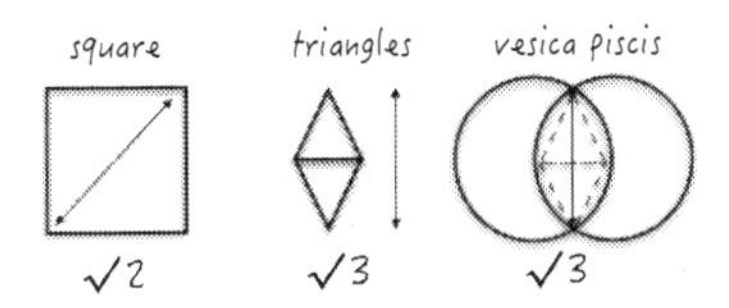

The Phi ratio is associated with the pentagon. If the exterior sides measure 1, the interior lines of a five-pointed star will equal phi.

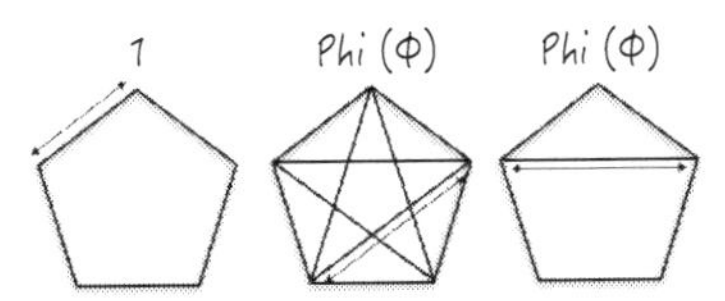

In three dimensional form, the Cube and the Octahedron nest through the ratio of $\sqrt{2}$ (*1*). The Cube and Dodecahedron unify through Phi (Φ) (*2*), whilst the Dodecahedron and Icosahedron nest through the ratio of $\sqrt{3}$ (*3*).

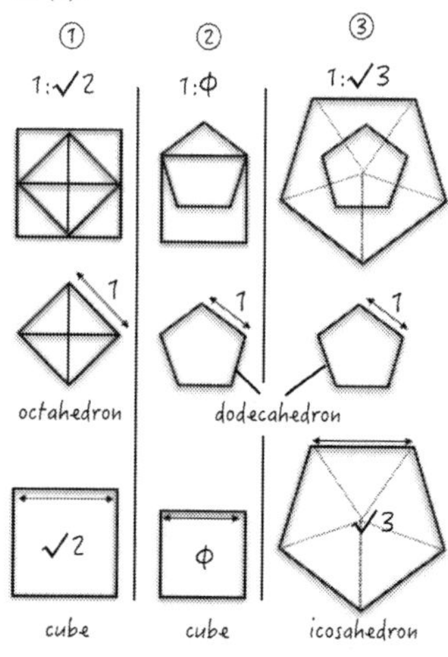

The Star-Tetrahedron nests inside the Cube (*1*). If the Cube has the side length of 1, the Star-Tetrahedron will measure $\sqrt{3}$ across opposite tips (*2*).

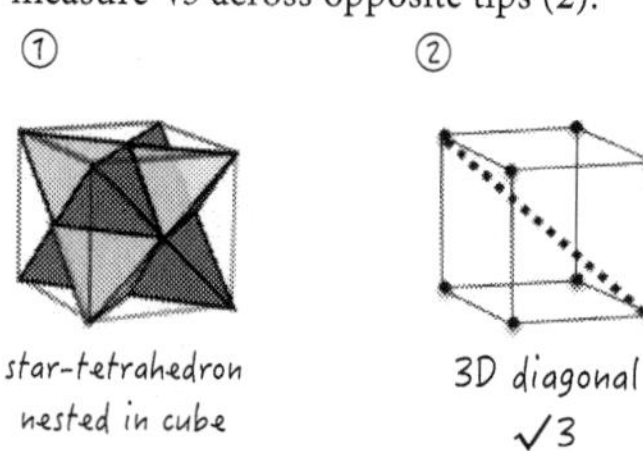

star-tetrahedron nested in cube

3D diagonal √3

PHI

If one is divided by a whole number, the result is its reciprocal value. All reciprocals exist between one and zero. Phi exhibits a curious relationship to its reciprocal. One divided by Phi, is exactly one less than its whole (Phi). Additionally, when Phi is squared, the result is exactly one greater. This gives Phi three different types of values:

1) Φ = greater Phi (1.6180...)

2) φ = lesser phi (0.6180...) ($1/\Phi$)

3) Φ^2 = Phi squared (2.618...)

We can demonstrate these qualities of Phi geometrically on a line that measures the length of Phi (1.6180). By subtracting 1, the remainder of the line is lesser phi.

Φ ├── 1 ──┼── 1/Φ ──┤

If the total length of the line equals Φ^2, subtracting 1 leaves the remainder Phi.

Φ² ├── Φ ──┼── 1 ──┤

Phi can also be calculated from $\sqrt{5}$:

$$\Phi = (\sqrt{5} \div 2) \pm 0.5$$

$\sqrt{5}$ divided by two equals $\sqrt{1.25}$. Add or subtract 0.5 to generate *greater* or *lesser* Phi. $\sqrt{0.25}$ is equal to 0.5. Therefore, phi can be complied from the addition or subtraction of two root numbers:

$$\Phi/\phi = \sqrt{1.25} \pm \sqrt{0.25}$$

PHI GEOMETRY

Previously, we created Phi from the triangle (*1*), and we presented the phi ratio in the pentagon (*2*). We can also generate phi from a square (*3*).

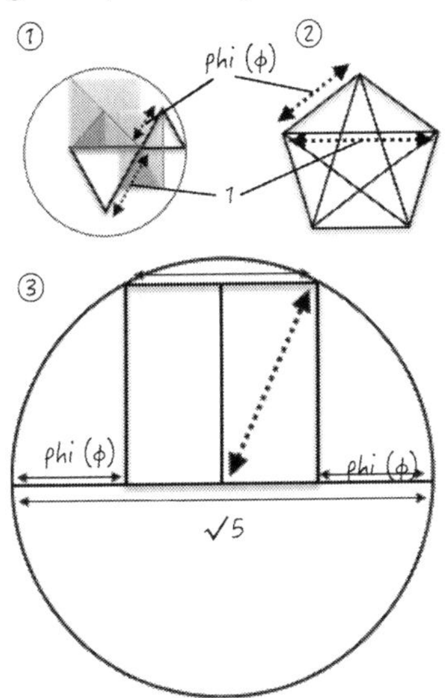

Divide the square into two rectangles. Position the compass on the midpoint of one side. Measure the distance to the opposite corner, and draw a circle. If the square has a side-length of 1, the diameter of the circle will be $\sqrt{5}$. Remove the square with a side of 1, and the remainder on either side is *lesser* phi (0.618). Phi evolves geometrically from three forms - the triangle, the square, and the pentagon - which are the only shapes used to construct the five Platonic Solids.

ROOT NUMBERS

Just as Phi has a specific relationship to its reciprocal value, the same is true for other root numbers.
$\sqrt{2} - 1$ is the reciprocal value of $\sqrt{2} + 1$.

$$\sqrt{2} = 1.4142$$
$$\sqrt{2} + 1 = 2.4142$$
$$\sqrt{2} - 1 = 0.4142$$

$$1 \div 2.4142 = 0.4142$$

Both numbers are found in this diagram.

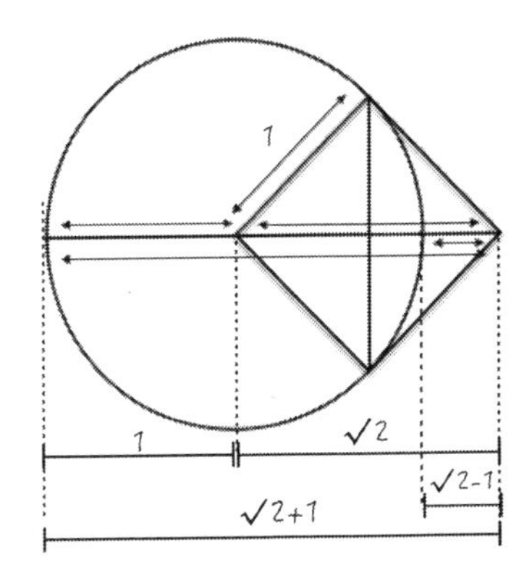

If the circle has a radius of 1, it cuts through the diagonal of the square to produce $\sqrt{2} - 1$. The radius of the circle combined with the diagonal of the square is $\sqrt{2} + 1$.

$\sqrt{3}$ has a similar relationship to its reciprocal. This time, the number 1 is replaced by the number 0.2679 ($2-\sqrt{3}$), which is the height of the Trion Re. We can define this number with the letter 'a' and make the following observations.

$\sqrt{3}$	= 1.7320
$\sqrt{3} + 1$	= 2.7320
$\sqrt{3} - 1$	= 0.7320
$a = 2 - \sqrt{3}$	= 0.2679

$$(\sqrt{3} - 1) \div a = \sqrt{3} + 1$$

The diagram below displays this numerical relationship.

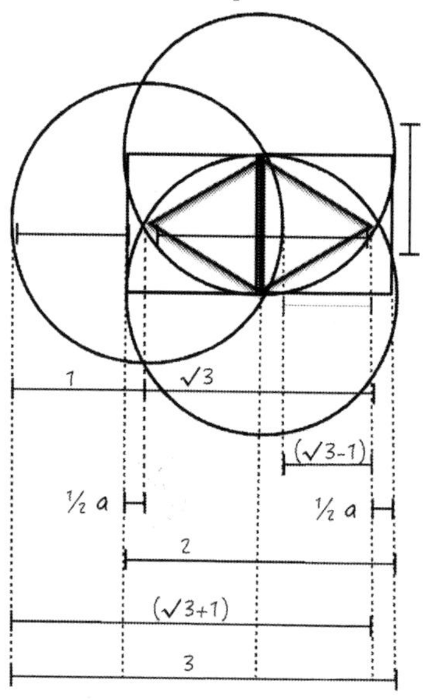

$\sqrt{5}$ displays a similar properties. This time, add or subtract 2 to find a value that matches its reciprocal.

$\sqrt{5}$	= 2.2360
$\sqrt{5} + 2$	= 4.2360
$\sqrt{5} - 2$	= 0.2360

$$1 \div (\sqrt{5} - 2) = \sqrt{5} + 2$$

The diagram below displays this numerical relationship.

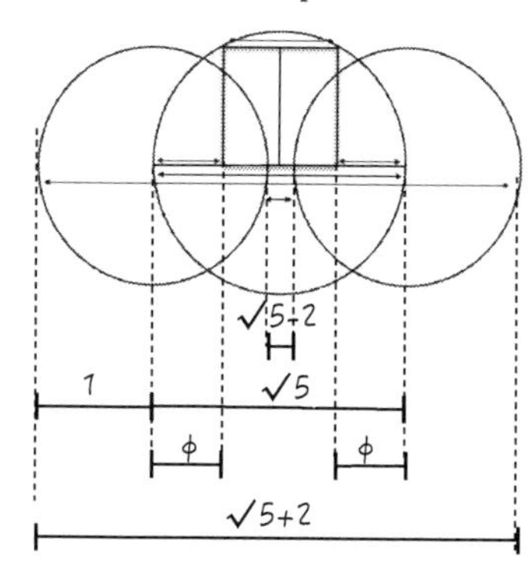

Pythagoras theorem is a mathematical principle found in all right-angle triangles.

$$a^2 + b^2 = c^2$$

The ratios $\sqrt{2}$, $\sqrt{3}$ and $\sqrt{5}$ can be placed into a right angle triangle based on this theorem.

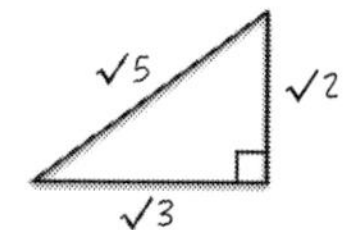

DUALS

Within the nested set of Platonic Solids are two sets of *Platonic Duals*.

SET 1		SET 2	
octahedron	cube	dodecahedr.	icosahedr.
6 corners 8 faces	8 corners 6 faces	20 corners 12 faces	12 corners 20 faces

The corner of each dual defines the centre face of its partner.
The Tetrahedron is a dual of itself. The corners of a smaller Tetrahedron, orientated at 180°, defines the centre face of the larger one.

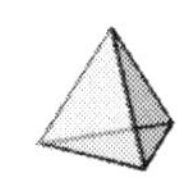
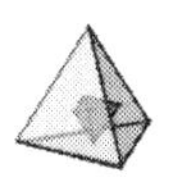
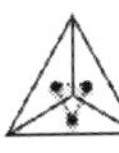

The characteristic of this interaction is utilised in the ratios of the nested Platonic Solid.

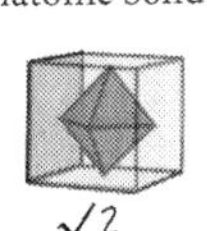

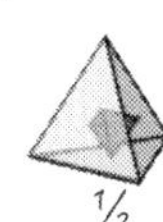

COMPOUND SOLIDS

Two Platonic Duals of equal size combine to create a 'Compound Solid'. The three types are; the Octahedron-Cube (*1*), two Tetrahedra that form the Star-Tetrahedron (*2*), and the Icosahedron-Dodecahedron (*3*).

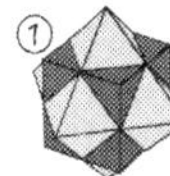

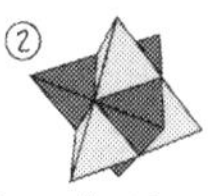

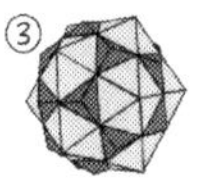

Two Dodecahedrons can be compounded around the Cube.

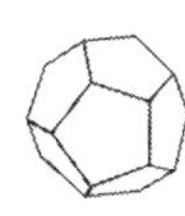

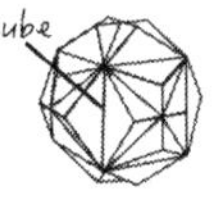

Internally, three rectangles (φ2 x φ) define the Dodecahedron (*page 48*). When rotated 90°, they form the compound of two Dodecahedra.

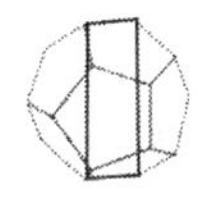
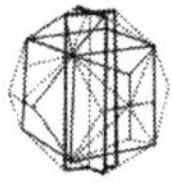
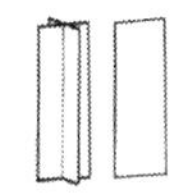

Thus, each face of the Cube appears with a cross over its centre. If the side of the Cube is 1, the cross is phi.

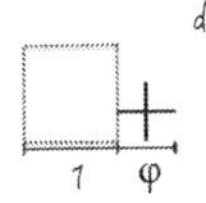

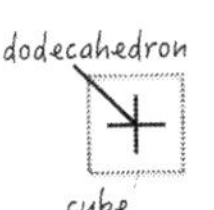

Inside the Cube nests two Tetrahedra and outside, two Dodecahedra.

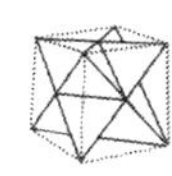
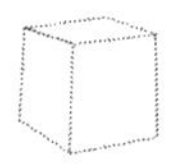
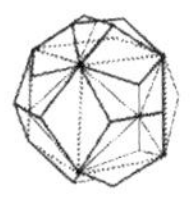

ORDER FROM CHAOS

The three main Compound Solids share a single unifying feature: a triangular face divided into four.

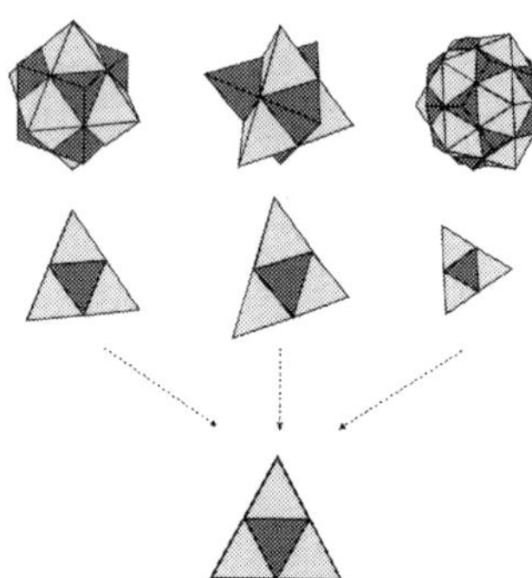

This image is generated by a random process. The 'Chaos Game' records the movement of a dot that follows a simple set of rules. The dot begins outside a triangle with corners defined, 1, 2 and 3 (*1*). The dot moves halfway on a trajectory towards one corner selected at random (*2*). The computer registers this new position and the process repeats indefinitely (*3*).

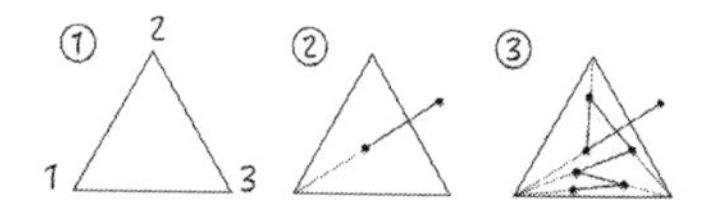

Over time, a pattern emerges. This fractal is called the Sierpinski Triangle.

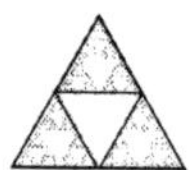

The Sierpinski Triangle has a specific form. It contains an up-facing (black) and a down-facing (white) triangle (*1*). Division occurs only within the black triangles, which generates the fractal. The white triangles never divide (*2*).

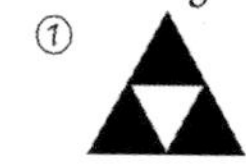

This process is similar to molecular behaviour (*1*). Billions of air molecules in a concert hall transmit sound perfectly through an apparently chaotic structure (*2*). Could this be ordered through such a simple blueprint?

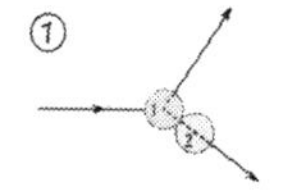

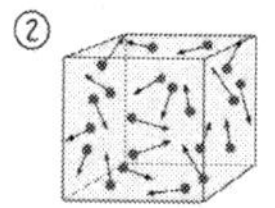

The ratio of a circle's circumference to its diameter (Pi) is generated randomly by Buffon's Needle Problem. Draw parallel lines that are the same distance apart as the length of the needles (*1*). Toss the needles onto the grid. Count how many cross the lines (black) (C) (*2*).

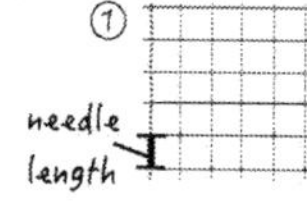

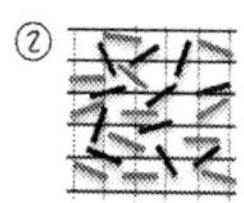

Repeat this and record the number of tosses (T). Divide T by half C. The greater the number of tosses, the closer the result approximates Pi (π).

$$T : (C/2) = \pi$$

Only two types of 2D plain exist (*page 11*). Buffon's needle problem defines Pi from a square, the Sierpinski fractal arises from the triangle. Both processes demonstrate an order out of chaos, negative entropy.

SPACE

A Cube fills an infinite space seamlessly. 8 corners make a Cube (*1*). A single point is surrounded by 8 Cubes (*2*). Atoms have a tendency to bond in sets of 8, called the *Octet Rule*.

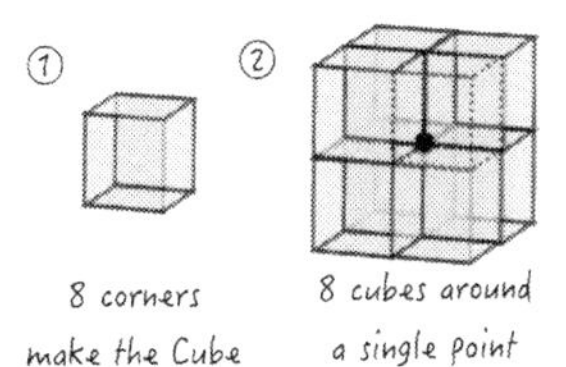

This can be surrounded by another 56, totalling a larger Cube of 64 (8^2). Element 56 marks the point on the periodic table, where the first f-orbitals begin to appear (*page 49*).

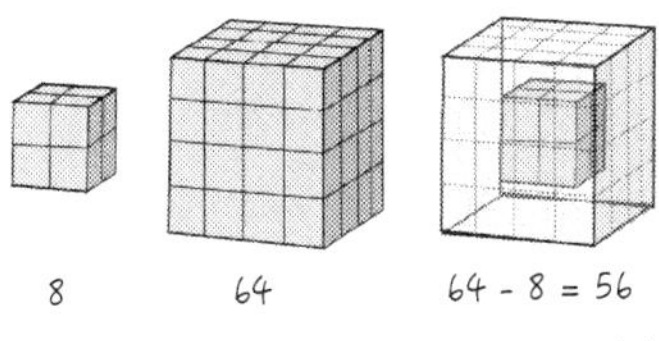

The Tetrahedron and Octahedron (*1*) combine to fill a three dimensional space in a similar way. 6 Octahedra surround a single point. This leaves gaps for 8 Tetrahedra (*2*). We call this quanta 'Octahedral Space'.

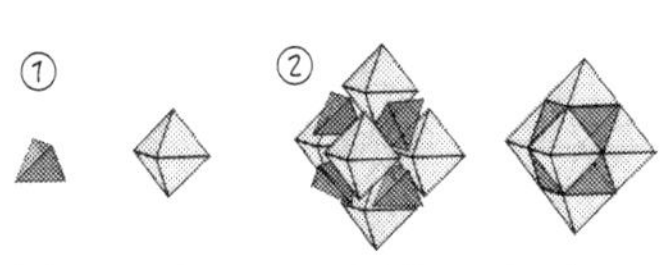

Expansion continues with 30 Octahedra and 40 Tetrahedra. A total of 84 polyhedra (48 Tetrahedra/36 Octahedra) create the next larger form (*1*). On the periodic table element 30 and 48 mark the points where the first and second set of d-orbitals complete (*page 49*). Element 84 is the first naturally radioactive element.
This structure comprises a total of 92 polyhedra. Element 92 (Uranium) is the last naturally occurring radioactive element. All elements beyond this are artificial and do not occur in nature. Octahedral Space defines many of the boundaries found in the periodic table.
4 small Tetrahedra and one Octahedron creates a larger Tetrahedra, double in size (*page 29*).
8 of these will fill the spaces left in the large Octahedron (*2*).
The fractal geometry of each face exactly matches the Sierpinski Triangle (*page 45*).

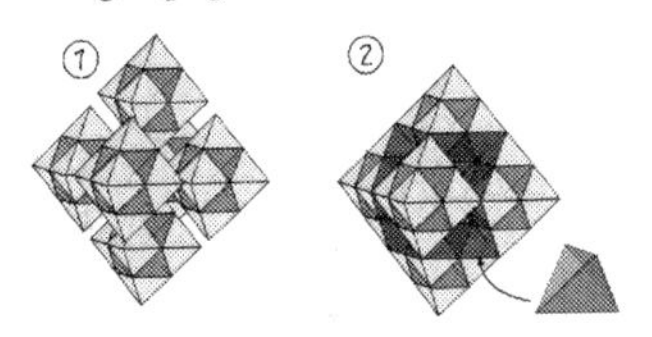

A square expands to fill an infinite 2D plane in the same way a Cube expands to fill 3D space (*1*). Hexagons and triangles work in harmony to fill a 2D plane (*2*), just like Tetrahedra and Octahedra fill 3D space. The square and the triangle are the only two forms to fill a 2D plane in this way (*page 11*).

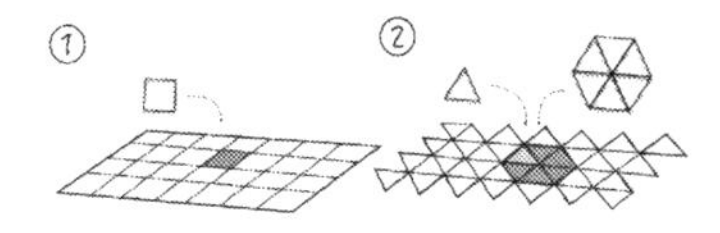

ARCHIMEDEAN SOLIDS

An Archimedean Solid is similar to a Platonic Solid. However, its faces consist of more than one shape. The Cuboctahedron combines 6 faces of the Cube, and 8 of the Octahedron. It appears in a quanta of Octahedral Space. Remove the Octahedra (*1*). The remaining Tetrahedra define the Cuboctahedron (*2*).

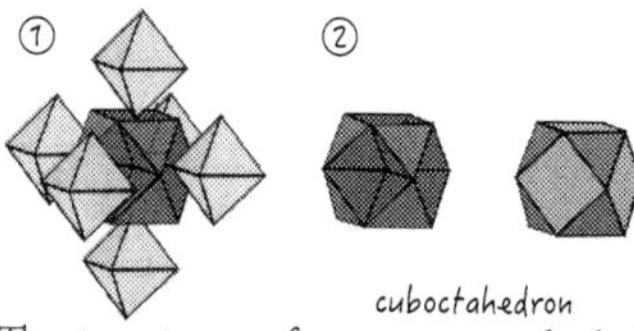

cuboctahedron

The two types of spaces - octahedral and tetrahedral - comprise opposite aspects of the Sierpinski Triangle. Octahedral Space produces the fractal geometry found in the right side-up triangle. Tetrahedral Space fills the upside-down triangle.

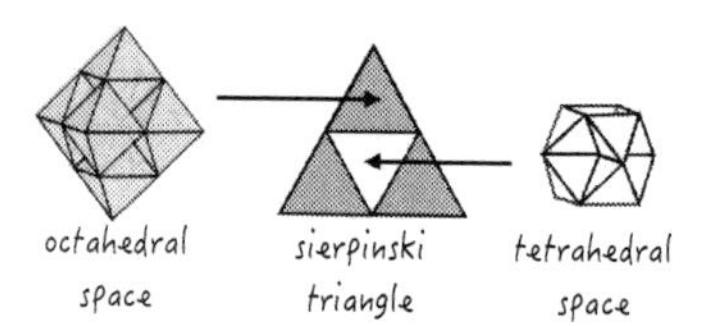

octahedral space — sierpinski triangle — tetrahedral space

12 spheres placed on each corner of the Cuboctahedron leaves space for another sphere in its centre. It is the perfect structure to compile spheres in 3D space.

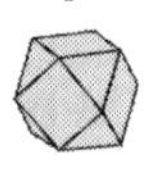
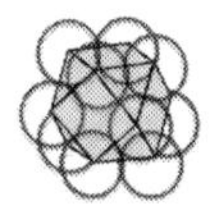

compilation of spheres in 3D

There are only 13 Archimedean Solids in existence. The Truncated Octahedron is the only one that fills 3D space uniformly by itself.

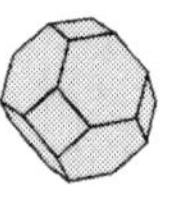

truncated octahedron

This form can fill three dimensional space through both, 'Cubic' and 'Octahedral' arrangements.

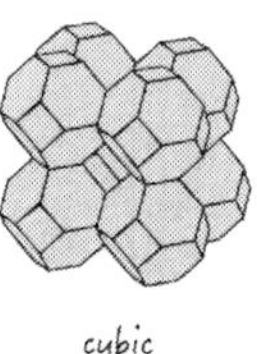

cubic structure

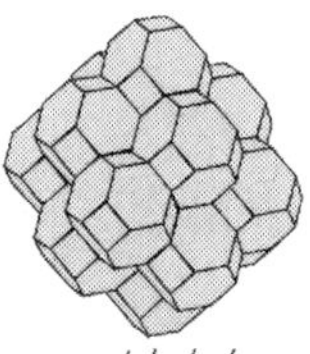

octahedral structure

The Truncated Octahedron looks similar to the Cuboctahedron. Both evolve out of the Octahedron. The Cuboctahedron is formed from an Octahedron with its edge-length divided into two. The Truncated Octahedron is derived from an Octahedron with sides divided into three.

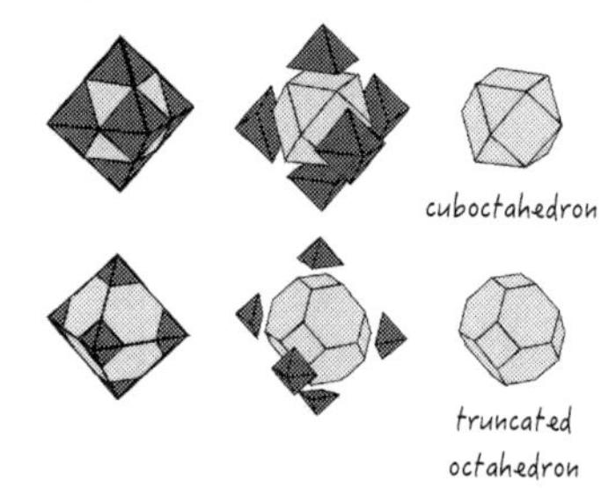

cuboctahedron

truncated octahedron

INVERSE GEOMETRY

In Metaphysics various shapes are seen from the perspective of their inverse: The dot is the inverse of a circle (*1*). The cross is the inverse of the square (*2*). The triangle is the inverse of the hexagon (*3*). The pentagon reflects into itself (*4*).

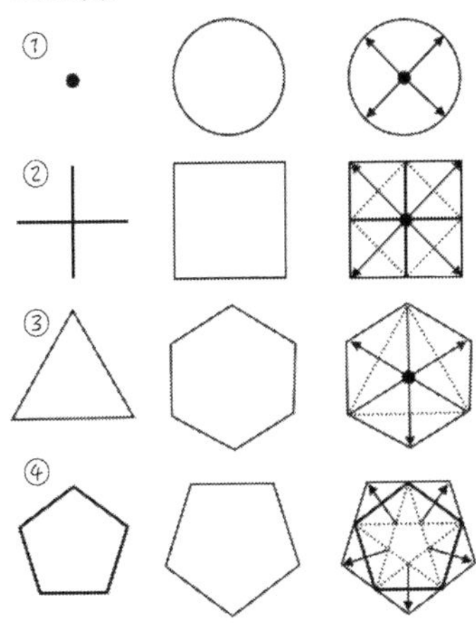

Inverse Geometry provides an interesting insight into the internal and external geometries of the Platonic Solids. The dot is the inverse of the sphere or torus. This movement is similar to those made by electromagnetic fields. Torus fields are found surrounding the earth, and many people believe in a human aura that circulates outside the body.

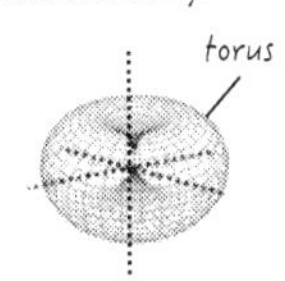

Three crossing lines define the internal geometry of a hexagon and also creates the inverse of the Octahedron (*1*). When lifted from 2D into 3D space, each angle shifts from 60° to 90°. From the Seed of Life, the Octahedron comes into existence (*2*). This presents a novel concept of a geometric method, which transforms 2D into 3D space.

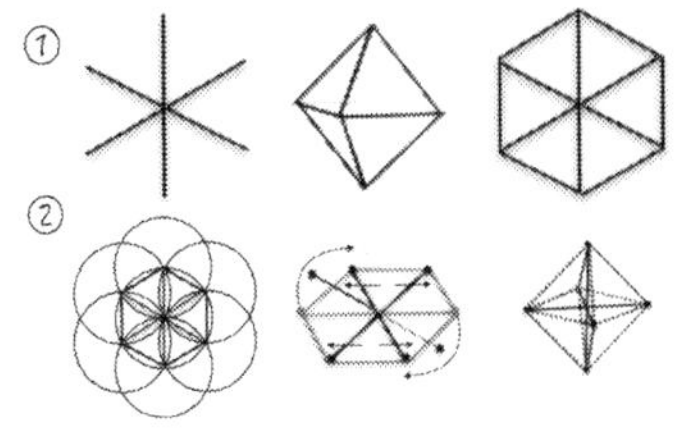

When the three lines are squared it forms the inverse of the Cube (*1*). Transforming these into phi rectangles forms the inverse of the Icosahedron (*2*), while rectangles with a ratio of 1:phi², create the inverse of a Dodecahedron (*3*). Its remaining 8 corners are defined by the Cube (*4*).

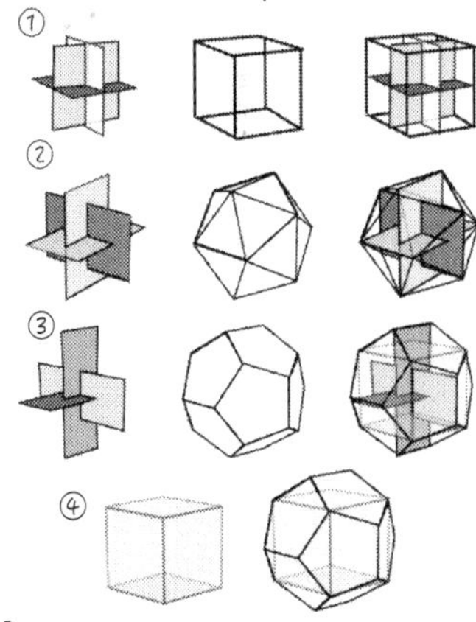

THE ATOM

The Octahedron/Tetrahedron (*1*), Cube (*2*), and Truncated Octahedron (*3*) are all *Space Filling Polyhedra*. Additionally, the Cuboctahedron (*4*) orders spheres into 3D space. All of these polyhedra consist only of triangular, square and hexagonal sides.

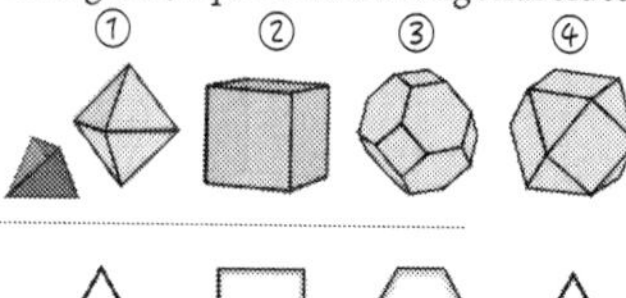

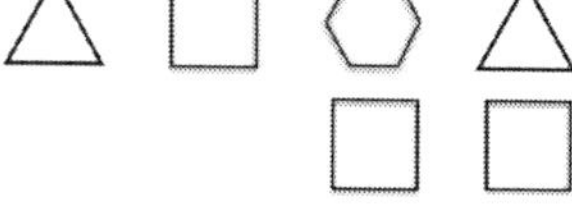

These shapes are the only forms that appear in the two types of regular 2D tessellation (*page 46*).
An Atom consists of a tiny nucleus surrounded by an electron cloud that is structured into different orbital shells. These orbitals appear in different configurations, which also display certain geometries:

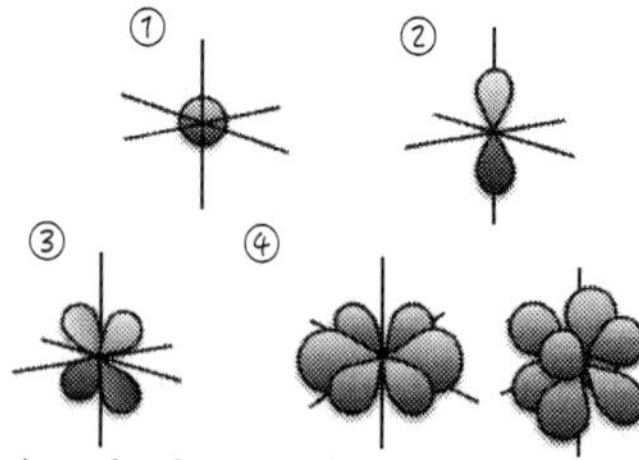

1) s-orbital/ circle
2) p-orbital/ line
3) d-orbital/ square
4) f-orbital/ hexagon and cube

Additionally, each of these orbital types are grouped together to form certain geometric formations, resembling the inverse geometry of the Space Filling Polyhedra.
P-orbitals come in three types, orientated on an x, y and z axis.

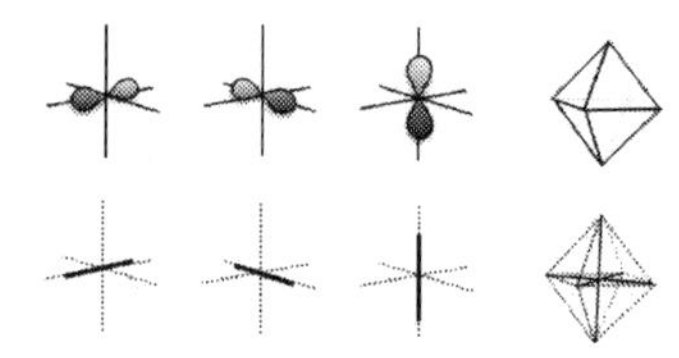

4 types of d-orbital consist of square planes. Three orientated to form the inverse geometry of a Cube and the fourth twisted at 45°.

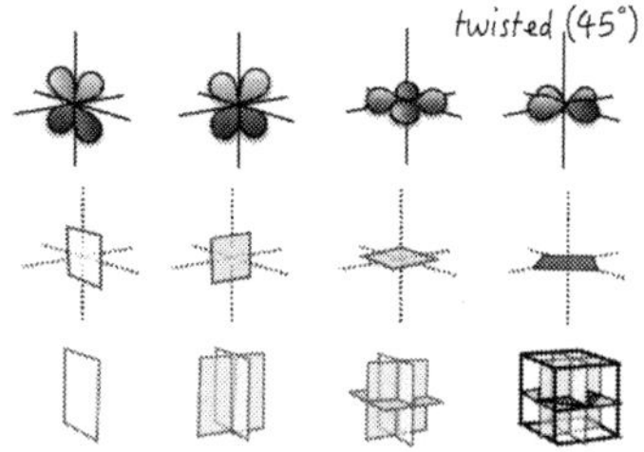

2 types of F-orbital appear in a cubic form (*1*) and another 4 are hexagonal, producing the internal geometry of the Truncated Octahedron (*2*).

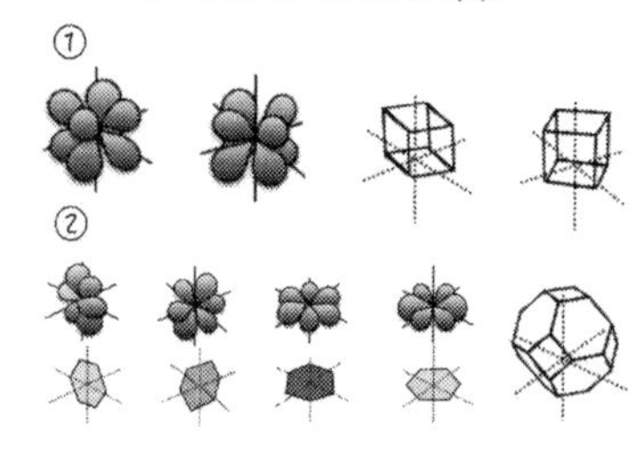

FIVE ELEMENTS

In Metaphysics, matter comes into existence through the enactment of the *Five Elements*. These elements are Fire, Water, Earth, Air and Aether (Electricity). We are unable to interact directly with Fire and Aether. The other three Elements form the states of matter we can touch: Liquid, Solid and Gas. This matches the model of the five nested Platonic Solids. Thus, we can ascribe an element to each one.

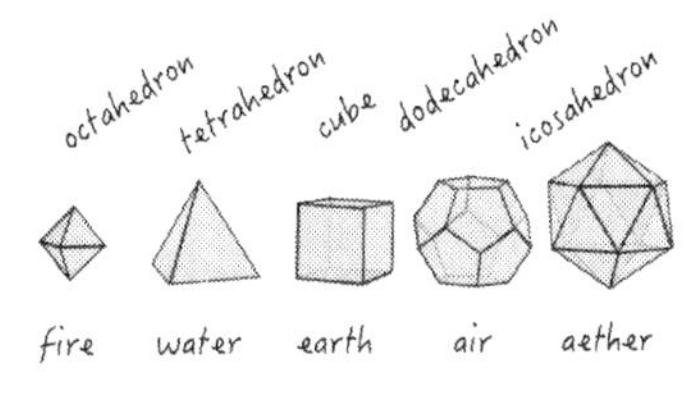

This view provides a metaphysical story of creation. When water falls onto Fire, it solidifies into rock, Earth. This represents the nesting of the Octahedron inside the Tetrahedron (*1*). Outside sits the Cube which represents the Earth that rises out of the water (*2*). Just as an island is a result of volcanic activity, it rises out of the ocean. Within the structure of the nested Solids, points expand from a unified centre. The Octahedron, the Dual of the Cube, defines the centre of each face of the Cube (*3*). The Tetrahedron directly interacts with the corners of the Cube (*4*). Fire and Water work together to create the orientation for the Cube.

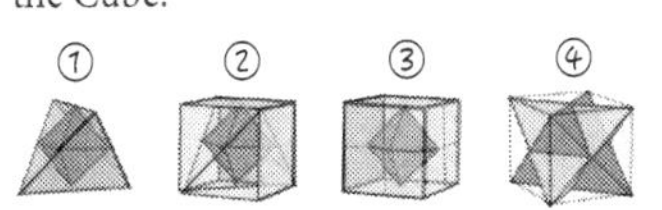

As the Water cools the Fire, it evaporates to become Air, the Dodecahedron. The Tetrahedron (Water) defines 8 corners of the Cube (Earth) which also interlink with 8 corners of the Dodecahedron (Air). Notice, oxygen is the 8th element in the periodic table.

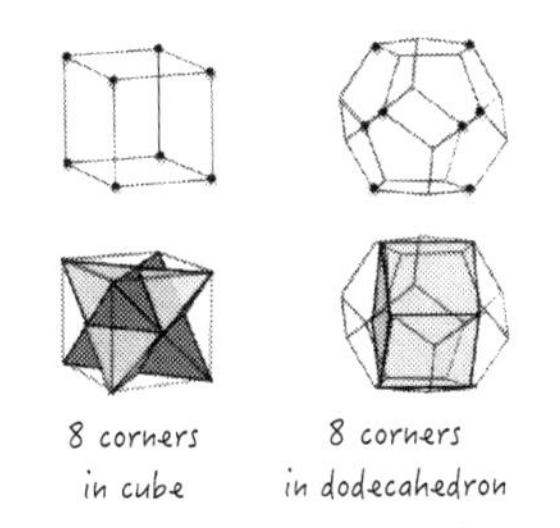

The Icosahedron (Electricity), envelopes the entire structure. The electricity ionises the air, causing it to rain back to Earth. Our planet consists of a molten centre of Fire, and an atmosphere made of Aether (electricity). In between the two shells, all life exists comprised of the three other elements.

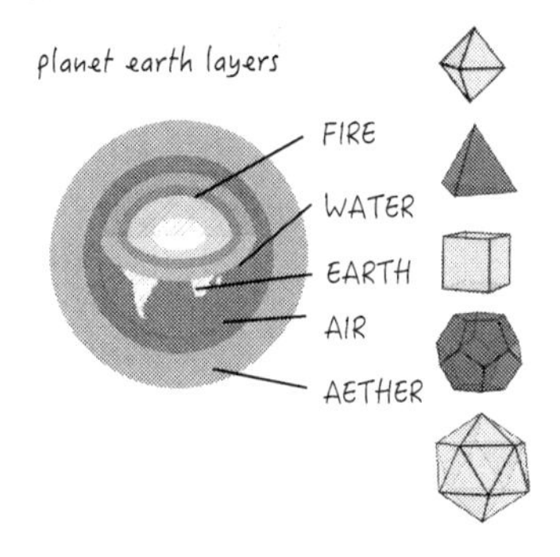

18 SOLIDS

This image below shows the relationships between the 5 Platonic Solids and the 13 Archimedean Solids. The 19th perfect solid is the Sphere.
There are three types of relationships: Truncations, Explosions and Turning.
Truncation: Remove the corners of one of the five Platonic Solids.
Explosion*: Move the faces away from the centre until they are separated by a side-length.
Turning: Remove the square faces of the Rhombicosidodecahedron and the Rhombicuboctahedron. Turn the structure until each of the square faces turns into 2 triangles.

1) Truncated Tetrahedron
2) Great Rhombicuboctahedron
3) Truncated Octahedron
4) Cuboctahedron
5) Truncated Cube
6) Rhombicuboctahedron
7) Rhombicosidodecahedron
8) Truncated Dodecahedron
9) Icosidodecahedron
10) Truncated Icosahedron
11) Rhombicosidodecahedron
12) Snub Cube
13) Snub Dodecahedron

truncation
tetrahedron
1

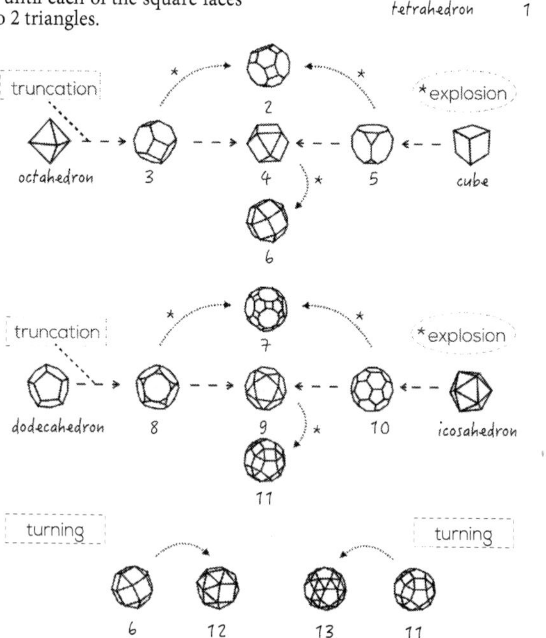

DATA TABLE

Solid	face	faces	vertices	edges	sphere diameter	dual	nesting order	side length ratio
Octahedron	triangle	8	6	12	1	Cube	1	1
Tetrahedron	triangle	4	4	6	$\sqrt{3}$	Tetrahedron	2	2
Cube	square	6	8	12	$\sqrt{3}$	Octahedron	3	$\sqrt{2}$
Dodecahedron	pentagon	12	20	30	$\sqrt{3}$	Icosahedron	4	$\varphi\sqrt{2}$
Icosahedron	triangle	20	12	30	$\sqrt{(\varphi^2 + 1)}$	Dodecahedron	5	$\varphi\sqrt{2}\sqrt{3}$

Solid	faces	vertices	edges
Cuboctahedron	14 faces: 6 squares + 8 triangles	12	24
Truncated Octahedron	14 faces: 8 hexagons + 6 squares	24	36

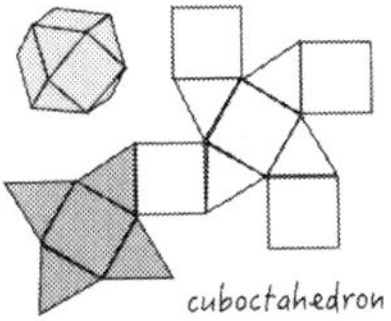

cuboctahedron

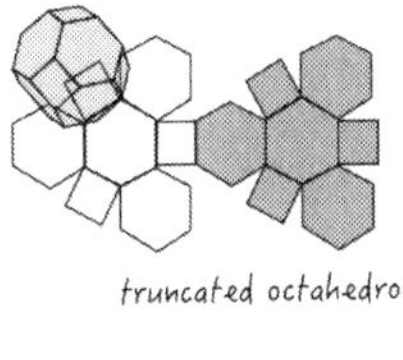

truncated octahedron

If you have read everything carefully, and found it a little bit interesting, then hold on to your hat, we're about to go interstellar!

More books from In 2 Infinity:

HANDBOOKS

The Template – A Journey through 3D
Discover the wonders of the five platonic solids, and how to create them.

13 Sacred Numbers – An Evolution of Consciousness
Turn a line into different equal sided shapes and uncover their spiritual metaphor. (*coming soon*)

The Ordinance – A Metaphysical Guide
Expansive and contractive Sacred Geometry describes the underlying mathematical structure of life. (*coming soon*)

BOOKS

In 2 Infinity – The Trilogy (*coming soon*)

Book 1 – Space and the Aether
If the Aether existed, it would fundamentally shift our perception of the atom, and the foundations of quantum theory itself.

Book 2 – Time and the Ordinance
Between viewing time as a linear or cyclic function, is there a middle way that could explain the unsolved mystery of gravity?

Book 3 – Numbers and the Code
Is there a secret code hidden within numbers, that could explain the meaning of life and consciousness?

INTWO
Infinity

www.intwoinfinity.com

23087967R00035

Printed in Great Britain
by Amazon